Cross-Cultural Scientific Exchanges in the Eastern Mediterranean, 1560–1660

Cross-Cultural Scientific Exchanges in the Eastern Mediterranean, 1560–1660

AVNER BEN-ZAKEN

The Johns Hopkins University Press
Baltimore

Printed in the United States of America on acid-free paper
2 4 6 8 9 7 5 3 1

The Johns Hopkins University Press
2715 North Charles Street
Baltimore, Maryland 21218-4363
www.press.jhu.edu

Library of Congress-Cataloging-in-Publication Data

Ben-Zaken, Avner.
Cross-cultural scientific exchanges in the eastern Mediterranean, 1560–1660 /
Avner Ben-Zaken.
p. cm.
Includes bibliographical references and index.
ISBN-13: 978-0-8018-9476-3 (hardcover : alk. paper)
ISBN-10: 0-8018-9476-X (hardcover : alk. paper)
1. Science—Mediterranean Region—History—16th century. 2. Science—Mediterranean Region—History—17th century. 3. Cultural exchange. I. Title.
Q127.M38B46 2010
509.182'209031—dc22 2009033025

A catalog record for this book is available from the British Library.

To my parents,
Moshe and Aliza Ben-Zaken

I have serious reasons to believe that the planet the little prince came from is Asteroid B-612. This asteroid has been sighted only once by telescope, by a Turkish astronomer, who had then made a formal demonstration of his discovery at an International Astronomical Congress. But no one had believed him on account of the way he was dressed. Grown-ups are like that . . . The astronomer repeated his demonstration, wearing a very elegant suit. And this time everyone believed him.

—Antoine de Saint-Exupéry, *The Little Prince*

When we are with one another in public, our everyday concern does not encounter just equipment and work; it likewise encounters what is "given" along with these: "affairs," understanding, incidents, mishaps. The "world" belongs to everyday trade and traffic as the soil from which they grow and the arena where they are displayed.

—Martin Heidegger, *Being and Time*

CONTENTS

Incommensurable Cultures?

IN 1660, A HUNGARIAN CONVERT and Ottoman bureaucrat, Ibrāhīm Efendi al-Zigetvari, found a Latin work that mentioned the Copernican cosmology, and, independently, he translated it into Arabic; only after completing the project did he show it to the *munajjim bāshī* (the sultan's chief astronomer/astrologer). At first, the chief astronomer did not approve of it, saying: "Europeans have many vanities similar to this one." But when al-Zigetvari prepared an almanac based on the astronomical tables of the Latin work, the chief astronomer saw that it conformed to the authoritative tables prepared by late medieval Muslim astronomers, and he became convinced of the value of the Latin work.

This story captures the various ingredients—personal trust, credibility of writers, utility, and chance exchange—that characterize the circulation of post-Copernican astronomy from West to East and encapsulate the shift in cultural self-consciousness of science. By the mid-seventeenth century, Europeans and Near Eastern practitioners became increasingly convinced of the shift of superiority in natural philosophy from the Near East to Europe. The concluding note to the full acceptance of the heliocentric cosmology in the Eastern Mediterranean came in the early nineteenth century, when an Ottoman-Turkish translation of a treatise of Edmund Halley (1656–1742), *Astronomiae cometicae synopsis* (1705), appeared in Istanbul, presenting the heliocentric cosmology as the exclusive cosmological system.[1] The two incidents provide only the epilogue to a century (1560–1660) of cross-cultural exchanges that are the subject of this book and that, on various grounds, were almost completely overlooked by contemporary scholarship.

For centuries, students of European and Near Eastern cultures have supposed that the two cultures developed along separate, linear paths and that once the birth of the Copernican cosmology inaugurated the "Scientific Revolution," modern scientific knowledge flowed in one direction—from West to East. This presumption is nourished by received historical narratives and by various craft-related constraints, particularly the culturally focused training of graduate students and the fixed cultural division in history departments.

Ingrained historical narratives sustained the belief that European and Eastern Mediterranean cultures have developed along separate paths. Early modern European intellectual history started, arguably, with the fall of Constantinople in 1453, when Greek texts flowed into Europe, giving Europeans access to classical texts previously known only through Arabic and Hebrew translations. From the fifteenth century onward, European intellectual culture followed the humanist vision of the Renaissance, which called for a repudiation of translations and a return to the original Greek and Latin texts. Scholars promptly embarked on retranslations of classical works. The fifteenth-century astronomer George Peurbach dismissed the Arabic translations, working with his student Regiomontanus to translate and publish Ptolemy's *Almagest* from the Greek. And the sixteenth-century mathematician Federico Commandino translated Euclid's *Elements* from Greek to Latin.

Concurrently, Islamic intellectual culture developed along a divergent course. After the medieval translation boom of Greek philosophy into Arabic, scholars wrote commentaries on the Arabic texts, and then commentaries on the commentaries. These second-generation commentaries were revered as ultimate authority. From the ninth through the fifteenth centuries, Islamic natural philosophy cultivated key questions about cosmology, including an increasingly pressing doubt about the Ptolemaic system. By the sixteenth century, Islamic scholars perceived themselves as much more advanced, considering European science backward, if not completely irrelevant.

With the appearance of the heliocentric Copernican cosmology, the seemingly intellectual incommensurability went beyond historical narratives. By the early 1600s, the two rival cultures had also presented conflicting cosmological paradigms. Astronomers in the Eastern Mediterranean cultivated and carried on the Ptolemaic geocentric tradition, while their European counterparts started promoting the revolutionary Copernican heliocentric cosmology. The shift from the geocentric to the heliocentric cosmology, according to European historians, anticipated experimentalism and Newtonian physics by commencing a process in which natural philosophers were torn from the "past" (their source of knowledge about nature) and moved toward a strongly held new preference for evidence derived from mathematics, experiment, and observation. Historians of Islamic science, for their part, chose to focus on the medieval golden age—a high point that led teleologically to a long, steep slide—and overlooked early modern

Arabic science, which seemed to lie passively in the shadow of the European "Scientific Revolution." As a result, the East had little role to play in the history of early modern science. At any rate, scientific cultures, apparently, became incommensurable.

Methodological constraints, in their turn, have further deepened the cultural chasm. The triumph of cultural history has inadvertently produced incommensurable cultures. Current cultural historians tend to see natural philosophers, with their ideas, instruments, practices, and objects, as culturally embedded. Numerous cultural accounts have successfully erased the constructed lines between culture and science, but at the same time have, unconsciously, underscored the dividing lines between adjacent, local cultures. As a side effect, cultural history narrowed the historical narrative to one of incommunicative subcultures.

The conventional training of historians adds further obstacles. Scholars have mostly focused on individual regions, each with its local languages and particular cultural practices. Through such specialized and restricted training, for many young scholars, their acquired skills dictated rigorous confines to their professional development. Constructed cultural categories became the norm, while networks of communication between cultures fell by the wayside. Cross-cultural exchanges, thus, slipped off the historiographic radar, and research still concentrates on one cultural field—although it is trivially accepted by now that no culture, surely no scientific culture, is an island.

With the expansion of empires, establishment of trading companies in the East, and extension of print culture, early modern intellectual practices went beyond monastic contemplation in a closed room. Scholars extensively exchanged books, manuscripts, letters, and instruments, and traveled to other intellectual centers to study and to historical sites and exceptional geographic locations to do further research. Merely an exchange of material objects could set off unintended and uncontrolled processes of circulation, such that scientific ideas and practices traversed cultural boundaries. The prologue to the expansion of early modern science was the circulation of Copernican cosmology in the Eastern Mediterranean, and as such, this is the quintessential scholarly site for expounding the ways in which science transcended entrenched practices and traveled across cultures. In exploring such circulations and exchanges, this book aims to stimulate a larger historical discussion, one with further-reaching implications: can independent scientific development really occur? If not, how does science travel between and propagate through cultures? How do scientific cultures communicate? What criteria are necessary for the cross-cultural circulation of natural philosophy?

In the past two decades, sociologists and historians of science have argued that trust must exist in the exchange process; they suggest that networks of trust permit the spread of scientific ideas and practices. Trust binds scientific subcultures together and is an essential condition for the production of consensually accepted knowledge. Accounting for scientific circulation between hostile cultures, however, intensifies the problems. How, then, can one recount a dialogue between practitioners living during transitional periods who could hardly carry on a conversation, much less build up networks of trust?

One way to resolve such historiographic problems is to shift the scholarly focus from a single cultural site to networks of connections between cultures. As astronomers strived to replicate their observations through corroborating observational data or by relying on astronomical tables from other locales, they struggled to extend their scientific interests to adjacent cultures and to earlier astronomical tables. Exchanges of astronomical knowledge, books, instruments, and practices, therefore, serve as a good scholarly locus for examining how, even without formal social ties, post-Copernican cosmologies still managed to move between vastly different cultures and to extend European astronomical projects eastward. Thus, the focus on practices of exchange overcomes the constructed cultural and paradigmatic boundaries and brings about a widely prismatic yet deeply contextual history of early modern science.

This book, then, offers a cross-cultural account of early modern science. It explores how the heliocentric cosmology of Nicolaus Copernicus, as well as the semi-heliocentric model of Tycho Brahe, circulated eastward. It sketches out pursuits of post-Copernican ideas, persons, instruments, and texts as they traveled and were translated, transformed, adopted, and exchanged across European and Eastern Mediterranean communities. The project emerged during graduate school, when I decided to further pursue an earlier search for evidence of the circulation of the Copernican cosmology in the Eastern Mediterranean. The topic has thus far been misevaluated or simply ignored, partly because of the traditional argument that the Copernican Revolution, which commenced the Scientific Revolution, played as a "key moment" that set off divergent paths of development—Europe "progressing" and the rest of the world "declining."[2] As the "dawn of a new historical period," the Scientific Revolution represented not only a general break from medieval Aristotelian natural philosophy or "disenchantment" with popular, "irrational" beliefs in the occult, but also a particular break from the Eastern Mediterranean, identified as the swamp of superstition and the home of magical practices.[3] Given these descriptions, it crossed my mind that an understanding of the shift in the hegemony in natural philosophy from the

Muslim world to Europe had a great potential for revealing the historical roots of the cultural tension between Western and Eastern Mediterranean societies.

The book took shape while I was searching for evidence on the spread of Copernican cosmology to the Eastern Mediterranean, working on the assumption that the divergence between the two cultures was not as clear as has been supposed. Several primary sources bring to light that the controversies over cosmology were molded through negotiations with Eastern Mediterranean cultures. Such evidence can be found in several extraordinary natural philosophy collections. In the Eastern Mediterranean, these include Dār al-Kuttub in Cairo and various archives in Istanbul, specifically those at the Suleymaniye library and the Topkapi museum, which have early modern astronomical works translated from Latin into the local languages of Arabic and Persian. And in European archives, particularly the Vatican Library, the Lincei Archive in Rome, and Biblioteca Comunale Teresiana in Mantua, as well as in early modern collections in the Huntington Library and the Clark Library in Los Angeles, there are a few items of natural philosophy and astronomy that refer to Eastern sources and are bilingual editions, or even have Arabic, Persian, or Hebrew marginalia. The Institute of Hebrew Manuscripts in Jerusalem contains Hebrew translations of Latin works made in Salonika or Istanbul, translations that may reveal the mediating role of Sephardic Jews in bridging European and Eastern Mediterranean scientific cultures. The sources come in a variety of forms: manuscripts and letters that discuss European astronomical works and commentaries (especially commentaries in introductions to longer works) or contain translations into Arabic and Persian; traveler's books and letters written in Latin or Hebrew; and instruments and objects of gift exchange.

In searching these collections, I found a few sources that provided information on the circulation of the new Copernican cosmology: a miniature depicting the scientific activities of the Ottoman Observatory during the 1570s; a Latin-Persian translation of a work on the Tychonic cosmology, inscribed in a letter from an Italian traveler to a fellow, Persian astronomer; a Hebrew book on cosmology published by a student of Galileo who traveled in the Eastern Mediterranean; a Latin-Persian dictionary of astronomy composed by an Oxford professor; and an Arabic translation of a Latin work of cosmology.

The global dispersal and diversity of the sources indicates neither a translation movement nor cross-cultural exchanges that were motivated by a strong intellectual agenda. Written in Arabic, Latin, Persian, and Hebrew, the intellectual written languages of the geographic region between Spain and Persia at that time, the sources reveal the existence of a nexus of places rather than a single center of circulation. Books, manuscripts, letters, and instruments in Latin, Arabic, Persian, Hebrew, and Italian, on which the new cosmologies rode, extended across social networks that had not one but several intellectual centers: Istanbul, Cairo, Venice, Lār, London, Alexandria, Naples, Goa, Paris, Baghdad, Crete, and the Holy Land. The net encompassed a broad area, catching a variety of practitioners in a variety of locations in a vast, overlapping European–Near Eastern space. Copernican cosmology did not simply emanate from a European center. Rather, chaotic cross-cultural flows engendered critical views about cosmology that became a source not only for Copernicus himself,[4] but also for the next generation of practitioners who carried on astronomical exchanges and travels in the century (1560–1660) after his death. Thus, the important story of the circulation of post-Copernican astronomy can best be told by tracing the complicated paths of these artifacts as they moved between West and East. Unfolding of the story of this most unsystematic circulation provides a cross-cultural reevaluation of the "Scientific Revolution."

The scope of the project is, by necessity, broad: it requires a methodology crossing both linguistic and geographic boundaries. More importantly, it involves the apparently paradoxical use of both micro-history and a global perspective. On the one hand, micro-history allows the recovery of local meanings and contexts. On the other hand, a global perspective that looks at an array of cultural locations exposes the interactions between the micro-cases.

The painstaking recovery of evidence and the illuminating corroboration of narratives rely on the assumption that history is not only what is reflected in documents but also what was lost in the cracks between them. The role of the historian is not only to recover sources and to synthesize them into a coherent narrative but also to take a bold step in attempting to recover what was lost in the cracks between the documents. Deductive logic and intelligent deductions fill in the cracks and bring about a thicker historical narrative.

The research questions addressed here concern the material aspects and the mechanics of cross-cultural exchanges. Who made the artifacts or pictures? Who printed the books, and how? How were they transmitted to other places, and why? How did translators encounter books? Who financed travels, and why? What sources did writers use, and how were they obtained? This book, then, is a journal of research, a detective-like narrative that focuses on textual metaphors, titles, and peculiarities that can be deciphered only from within the context of the writers. By digging into these peculiarities, beyond the surface of the text, we can reveal their subtext, invoke their deeper meaning. Circulation of the universal values of the new cosmologies, when placed in such webs of local meanings, brings to light historical complexities that go beyond conventional historical explanations.

Science, like any other cultural production, consists of exchanging, altering, and borrowing from adjacent cultures and earlier time periods and is a product of socially driven networks that connect intellectual centers.[5] Practices of crossings and exchanges, placed at the heart of early modern scientific and cultural production, played a central role in resolving the early seventeenth-century cosmological controversies. Such cross-cultural practices cropped up in the rich and stimulating cultural space of the Eastern Mediterranean, where the margins of European culture overlapped the margins of Near Eastern culture to create a quintessential "zone of mutual embrace." Social networks entangled cultural margins and allowed travelers, incidental buyers and traders, diplomats and bureaucrats, and pirates and captives to cross paradigmatic and cultural boundaries and exchange objects and knowledge about the new Copernican and Tychonic cosmologies. As this book shows, various cross-cultural exchanges transpired in the "mutually embraced zone" of the Eastern Mediterranean, extending post-Copernican cosmologies eastward, where they were negotiated, shaped, and resolved.

Taking account of distantly scattered exchanges represents more than just the presentation of a sequence of various cultural contexts. Common values and procedures guided the practitioners in their quest to resolve the cosmological controversy. In the backdrop to such exchanges, the myth of a true cosmology already stated in the past bloomed and served as an overarching frame of reference. For cross-cultural practitioners of astronomy, science remained, more than anything, a historical endeavor.

Each chapter explores, in a micro-historical mode, a marginal textual object, written mostly by marginal figures in the history of science. Looking at things marginal sheds new light on the "Scientific Revolution," showing that early modern science owed its formation not exclusively to "monadic" cultures in the traditional intellectual centers, but also to vibrant, everyday cross-cultural exchanges on the margins of cultural fields.

Trading Clocks, Globes, and Captives in the End Time

IN 1580, AN ANONYMOUS PAINTER illustrated a manuscript entitled *Shāhinshāhnāma* (Book of the King of Kings). One of his miniatures documents scientific activity at the observatory established by the newly crowned Ottoman sultan, Murād III. When he came to power in 1574, he urged an Egyptian judge and rising star in natural science, Taqī al-Dīn Muḥammad Ibn-Ma'ārūf (1526–85), to come to his court and build the observatory. With exceptional knowledge in the mechanical arts, Taqī al-Dīn constructed instruments and built mechanical clocks for an observation of the comet of 1577. In the same decade, Tycho Brahe settled at Uraniborg, in Hveen, Denmark, from where he observed the same comet and made astronomical observations until the end of the century. The last observatory of Islam and the first significant observatory of Europe coexisted for many years. Current historiography tends to present the two projects as developing along separate paths. Yet, if we examine closely the details in the *Shāhinshāhnāma* miniature, we can extract clues about a possible connection between Taqī al-Dīn's and Tycho's worlds.

Astronomers played many roles at court. They relieved anxieties of rulers they served and helped provide them with displays of power and influence. During wars, times of crisis, or natural disasters, astronomers were summoned to lengthy and usually tense meetings in which they looked for guidance from the stars. They also attended solemn receptions, where ambassadors from far-flung nations brought with them gifts of scientific objects and books. More than any other scientific figures of the sixteenth century, astronomers operated within cross-cultural networks and were aware of adjacent cultures. Traditional historiography, however, tends to view them through a narrow cultural lens, cutting them off from their adjoining cultures. Such was the case with the narratives about Tycho Brahe and Taqī al-Dīn.

Relying on Pierre Gassendi's presentation of Tycho,[1] John Louis Dreyer's 1890 account gave us Tycho as an architect of new ideas that gave birth to an entire generation of astronomers. Dreyer looked for ways to prove that Tycho had cut himself off from traditional astronomy and presented him as the latest link in a cross-cultural chain that, in some cases, took in Islamic astronomy, especially that of al-Battānī (d. 929) and al-Zarqāll (d. 1087).[2] In the late 1980s, Victor Thoren shifted the discussion from the "history of ideas" to "intellectual history," focusing on the "person" Tycho, how his work changed society, and how he functioned within the Danish nobility.[3] In 2000, John Christianson presented Tycho's project as a political display of power and as embedded in contemporary culture, a presentation that blurred the lines used in previous historiography to define Tycho's work—the lines between culture, politics, and natural philosophy.[4] From Dreyer to Christianson, the result has been a purely European Tycho, detached historically from his Islamic predecessors and spatially from the Ottoman Empire—which in fact played an enormous role in shaping Tycho's broader European culture, politics, and natural philosophy.

In the historiography of Islamic science, Taqī al-Dīn emerges in only a slightly different manner. Scholars have tried to show that even up to the seventeenth century, natural philosophy in Islam was still viable, and they have looked to Taqī al-Dīn as the last representative of Islam's "golden age." They frame Taqī al-Dīn's achievements as an internal scientific product of an Islamic culture, operating without regard to Europe's new astronomy, natural philosophy, and natural history. In the 1950s, Sevim Tekeli elevated the significance of Taqī al-Dīn's work to that of a challenge to the Ptolemaic system. Moreover, she looked for similarities between Taqī al-Dīn and Tycho by comparing their astronomical instruments,[5] and framed a picture of Taqī al-Dīn as "the Tycho Brahe of the Ottoman Empire." Another Turkish scholar, Muammer Dizer, with a nationalist agenda, denied a possible connection between Taqī al-Dīn and the new astronomy and mechanics in Europe, wishing to present him as working concurrently—but without the taint of diffusion—on the same themes.[6] Moreover, some scholars have appreciated Taqī al-Dīn's achievements in terms of institutional history. Aydin Sayili, for example, produced a concise history of the Ottoman Observatory, the last observatory of Islam, its instruments and financial resources. However, he gave little attention to the contemporary cultural sentiments that prompted its establishment.[7]

Within the mutually embraced Euro-Ottoman cultural space, knowledge of nature was derived in disparate lands through parallel observations of common celestial phenomena and through the circulation of objects and people. However, the circulation of knowledge occurred and was expressed in terms of a cosmological contest, a contest inherent in apocalyptic visions that brought parallel astronomical projects into competition.

Clocks and Globes, Jews and Captives: Taqī Al-Dīn's Intellectual Sources

In looking for clues to Taqī al-Dīn's awareness of the rising, new natural philosophy in Europe, we must pay close attention not just to his scientific writings. Although most of his writings are technical scientific texts, their titles, introductions, and illustrations, in combination with texts by other writers in his surroundings (not necessarily in natural philosophy), give us access to the cultural crossroads where Taqī al-Dīn was situated—where cultural fields overlapped. There are more than a dozen scientific writings by Taqī al-Dīn, found in various libraries and archives in Istanbul, that document his intellectual interests; they vary from astronomical models of the planets and observational data of the comet of 1577 to manuscripts of mechanical instructions on how to build automata such as clocks and watermills. However, there is also a piece of cultural writing, a manuscript of poetry entitled *Shāhinshāhnāma* (Book of the King of Kings), which contains clues about the culture of the Ottoman Observatory. The work was composed in 1580 by the Persian poet 'Alā' al-Dīn al-Manṣūr. It obsequiously glorifies Sultan Murād III and was intended to be used as an object of gift exchange with European courts. The manuscript, illustrated with miniatures, also includes poems praising the Ottoman astronomer Taqī al-Dīn.[8]

We know that the Ottoman Observatory was demolished in 1580, for false astrological predictions and other problems, and, even before that, it was under pressure from the clergy, but the motives for its establishment and the scientific culture that allowed it to be built have not been explored.[9] Because its astrological aspects, mentioned in some European studies, are most often dismissed by modern Islamic scholars "for showing a strong tendency to look down" on Taqī al-Dīn's project, we do not have accounts that fully explore the contextual and utilitarian circumstances that led to establishment of the observatory.[10] Thus, the "history of ideas" type of scholarship on Taqī al-Dīn shows us an astronomy evolving internally, detached from political pressures for astrological predictions that revealed and resolved potential hazards to rulers. But we need to ask, what were the sources that helped Taqī al-Dīn fashion a persona as one who had morphed from an Egyptian mechanic into chief astronomer of the Ottoman

Empire? *Shāhinshāhnāma* supplies clues to the cultural currents underlying Taqī al-Dīn's scientific writings.

'Alā' al-Dīn al-Manṣūr's *Shāhinshāhnāma* is a chronicle, in Persian verse, that deals with the early years of the reign of Sultan Murād III (1574–95). It starts with poems praising God, the Prophet, and the coronation of Sultan Murād III. The rest consists of poems on miscellaneous events in history, among which is a poem about the Istanbul observatory, the comet of 1577, and the wars with Persia at the end of the 1570s. The poem implies political as well as scholarly motives for establishment of the observatory. Al-Manṣūr indicates that the old astronomical tables of Naṣīr al-Dīn al-Ṭūsī (d. 1274) and Ulugh Beg (d. 1449) "had become worn out" and insufficient for astrological predictions.

> The heavenly bodies were impatiently waiting for the observers,
> Just as the ascendants of persons of good fortune awaited
> The new astronomical tables.
> Then suddenly, with the splendor of the sovereign of the earth,
> The master of the rulers of the time,
> The highly placed and world-conquering Emperor,
> The Shāhinshāh of the climes, Sultan Murād, *things changed completely*.
> The surface of earth, with its wheel-like shape, is the roving-ground of this potentate . . .
> When he [Sultan Murād] issues orders for making observations and compiling astronomical tables
> The stars will descend and prostrate themselves before him,
> And when, in their endeavors, the astronomers are backed up by his sovereign power
> They will carry off the crown from the stars of the Ursa Minor.[11]

During the first phase of Sultan Murād III's reign, "things changed completely," and the work of the observatory aimed to move the horoscope of the sultan out of Ursa Minor to change his destiny. Noticeably, a fear of heavenly events prevailed. Moreover, the poem supplies clues regarding Taqī al-Dīn's skills in mathematics and mechanics.

> To the felicity of the seat of his victorious government, came from Cairo a qāḍī [judge] of high merits.
> His proficiency in mathematics went back to his forefathers.
> This man handles the pen with extreme swiftness and his name is Taqī al-Dīn.
> In the art of calculation the pen is servile and compliant in his hand.

With alacrity he fills the pages with numerals and figures.
He has surpassed Ibn Shāṭir [the noted fourteenth-century Islamic astronomer] and has taken his pre-eminence away from him.
In [Ptolemy's] the *Almagest* he has clarified many intricate parts,
And in Euclid's *Elements* he has disentangled many difficult points.
With the help of compasses and rulers and through strange figures he measured the latitudes and longitudes of all parts of the earth.[12]

Using poetic hyperbole, especially in describing Taqī al-Dīn's place in the history of science, the poem emphasizes his scholarship as rooted in the Islamic tradition of science. Although Taqī al-Dīn built many of the observatory's instruments, some were imported.[13] The incorporation of the latest European technologies, for the first time in the practice of astronomical observation in the Islamic world, left traces not only in texts but in graphic depictions. One of the *Shāhinshāhnāma* miniatures depicts astronomers working on an armillary sphere—a skeletal celestial sphere with a model of Earth at its center. Another shows the comet of 1577 in a manner reminiscent of European paintings of the same comet. The third miniature depicts the small observatory in which the staff observed and calculated data (figure 1); it may be seen as a snapshot of the actual activity in the Ottoman Observatory, and through it we can search for evidence for European connections. This particular miniature is significant not just because it can shed light on various broad issues, but because of the details at its margins. Apart from the fifteen staff members in three rows, we can identify a terrestrial globe (center of bottom row) and a mechanical clock (right margin of middle row). Posed so prominently in the miniature, the terrestrial globe implies a geocentric worldview. On it, we see Africa, Europe, the Mediterranean, and parts of Asia. Piri Reis, an Ottoman sea captain, is thought to have drawn a map of the Americas in the early sixteenth century, but the miniature does not give any indication that the globe was updated according to the new cartography.

Moving away for a moment from scholarly sources, to a 1991 brochure from Christie's Gallery in London, we see something interesting. The brochure offers for sale two globes—one celestial, the other terrestrial—together entitled "The Globes of Murād." They were constructed in Antwerp in 1579 and dedicated to Sultan Murād III. The Latin inscription on the celestial globe reads: "Amurat III, Suleyman, by the grace of the great god in the heavens, the only king of all the kingdoms of the world, the emperor [and] sultan of the Turks" (figure 2).[14] The Christie's brochure does not make any connection between Taqī al-Dīn's globe

Figure 1. Taqī al-Dīn and his staff in action. At the top, a Persian inscription describes the training of Taqī al-Dīn's staff: "And they also built a small-scale observatory in [the] vicinity of the main building / In it fifteen distinguished men of science / Were in readiness in the service of Taqī al-Dīn / In the Observations made with each instrument / Five wise learned men cooperated." From 'Alā' al-Dīn al-Manṣūr, *Shāhinshāhnāma* (ca. 1581), F 1404, p. 57a, Istanbul University Library.

and the Christie's terrestrial globe that appears to be identical to the globe in the miniature. The terrestrial globe was based on Gerard Mercator's map of 1541 and, of course, included the New World. The celestial globe's stellar positions, however, were based on the celestial map of Johannes Schöner, printed in the work *Opera mathematica* (published in 1551 and 1561), which relied on the astronomical tables in Copernicus's *De revolutionibus* of 1543.[15] Schöner was, in fact, one of the first to acknowledge Copernicus's achievements, having informed Rheticus—a mathematics professor of Melanchton's circle—of Copernicus's work.

Through the celestial globe, pieces of European astronomy lay hidden in the Ottoman Observatory. It may be that Taqī al-Dīn used the globe without knowing that its numerical data derived from new astronomical works. His interactions with the terrestrial globe and other instruments created a cultural self-consciousness within a unified worldwide space. We further learn about an encompassing perception of space from the poem quoted above: "With the help of compasses and rulers and through strange figures he measured the latitudes and longitudes of all parts of the earth." While previous Islamic observatories at Marāgha (thirteenth century) and Samarqand (fifteenth century) produced tables of longitudes and latitudes of Islamic cities, as well as of adjacent and traditional locales (Constantinople, Alexandria, and Rome), the Ottoman Observatory worked within a worldwide space, aiming to coordinate data that went beyond traditional locales.

The newer, wider spatial worldview determined the practice of observations and recordings for Taqī al-Dīn and his staff. It linked with a worldwide Ottoman Empire that grew not only because of its aspirations to conquer parts of Europe, but also as a result of a growing awareness of the discovery of the New World. In wishing to determine the longitudes and latitudes of the world, the Ottomans could pretend to conquer the world, at least through maps, and by so doing revealed their political anxiety over falling behind the Europeans in the race to the New World. The poem makes the connection between political aspirations to rule and astronomical techniques to improve data, and it presents Murād III as "world-conquering Emperor . . . The king of kings of climes." In the Latin inscription on the celestial globe (see figure 2), he appears as "the only king of all the kingdoms of the world," who wanted to effect radical changes and issued an order for "making observations and compiling astronomical tables." And so, Taqī al-Dīn was invited to "measure all parts of the world."

Figure 2. Inscription at the bottom of the celestial globe of Murād III: "Amurathes tertius Magni in coelo Dei gratia solymanus solus omnium regnum mundi rex imperator sulthanus Turcarum 1579." From Christie's brochure, *The Murad III Globes: The Property of a Lady to Be Offered as Lot 139 in a Sale of Valuable Travel and Natural History Books, Atlases, Maps, and Important Globes . . .* © Christie's Images Limited, 1991.

The Bānkām and the Changing Universal Time

The other scientific object of interest in the miniature (see figure 1) is the mechanical clock, placed on the right-hand side, corresponding to the center row of the observatory staff. A sharp eye will notice a spring-driven clock. No earlier observatory in the Islamic world incorporated a mechanical clock, and Taqī al-Dīn was a pioneer in discussing the philosophical implications of its use.

During the late sixteenth century, mechanical clocks, as well as other automata, were one of the main European currencies of gift exchange and trade. Carlo Cippola's account of clocks and culture supplies ample evidence for European clocks arriving in China through gift exchanges. Because the clock created fascination, especially at court, courtiers received it as a toy.[16]

Exchanges of automata and clocks worked similarly at the Ottoman court.

Many Italian technicians brought into the Ottoman Empire new trends in mechanics, but neither clock-making nor clocks were in wide use.[17] Otto Kurz provides numerous examples of the transmission of clocks from Europe to the Ottomans. Sultan Murād III, for instance, employed a prisoner-of-war clockmaker from Graz. And the famous Grand Vizier Mehmet Pasha Sokullu (d. 1579) employed a European captive as "clockmaker and steward." Moreover, in 1590, Rudolf II, Holy Roman Emperor, sent Sultan Murād III a watch and a clock.[18] Rudolf II's envoy, Sonnegk, who bore one of the gifts, described the reception of the clock, made especially for Murād III: "[It was] a clock in the shape of a castle, the gate of which opens at the stroke of the hour and out comes a figure of the Sultan on horseback followed by pashas, all of silver. After having made their round the cavalcade disappears behind another gate. Then the bells announce the hours, and everything is so pleasant and magnificent that the Christians were amazed and the Turks enchanted."[19] Such exchanges occurred through artisans, merchants, and diplomatic envoys to Istanbul who had clockmakers in their retinues. Most technicians returned to their native countries, but some did remain in the Ottoman court.[20] Ghiselin de Busbecq, a European envoy in Constantinople who presented mechanical clocks to the Ottomans, confirms the impression. In one of his *Turkish Letters* from 1555, de Busbecq describes how his watches were received in Istanbul and, in passing, criticizes the Ottoman technique of timekeeping. One of his colleagues made use of a watch and informed his other Ottoman friends at court "that it is nearly morning, or that the sun would not rise for some time, as the case might be. When they had once or twice proved the truth of his report, they trusted the watches implicitly, and expressed their admiration at their accuracy."[21]

Mechanical clocks at the Ottoman court were, indeed, a stylish form of high-level gift exchange; however, there was also a local production of clocks. In fact, Taqī al-Dīn was the first Islamic artisan that we know of to build a mechanical clock and was the first to name mechanical clocks in Arabic as *bānkāmāt*. His manuscript *The Revolving Planets and the Revolving Clocks* (*Kitāb al-kawākib al-durrīyah fī bānkāmāt al-dawrīyah*) depicts, stage by stage, with careful diagrams and instructions, the construction of a clock with second hands.[22] He argues that the use of such a clock would increase precision in astronomy and eventually would lead to the unfolding of the most refined secrets of nature. For Taqī al-Dīn, the building of the mechanical clock and its application to astronomy required special knowledge in hermeticism and strong intuitive perception.

The mechanical clock enabled Taqī al-Dīn to measure celestial motions precisely. Only a few years after clocks with second hands appeared in Europe, Taqī al-Dīn discussed the importance of minutes and seconds for astronomy. Measurement of the line of motion with minutes and seconds led to an analysis of geometry using new subdivisions, a process that shortened the road to "mechanical philosophy" and to perceptions of nature no longer completely based on integrated metaphysical assumptions. In one of his mathematical manuscripts, Taqī al-Dīn even discussed the way in which decimal fractions in trigonometry and astronomy could improve the results in planetary model building.[23]

Much has been written in the history of science about the way in which European mechanical clocks and automata replicated the perceived structure of the world, which in turn invoked a mechanical cosmology. Taqī al-Dīn, in a sense, participated in a larger trend to describe nature as a machine. In his other mechanical work from 1552, *Book of the Sublime Way of Spiritual Mechanics* (*Kitāb al-ṭuruq al-samiyyah fī al-ālāt al-rūḥānīyyah*),[24] he presents spiritual mechanics as the product of practical mastery in alchemy, talismans, and intuitive perception.[25] His mechanical abilities developed during his first visit to Istanbul in 1546, when he built "a rolling machine for shish-kebab."[26] He also gives diagrams of utilitarian machines such as a waterwheel and a "bed for lovers" (سرير العاشيق), in which a beautiful picture would jump up every time the person lying on the bed turned on his side.[27] But such mechanics did not end up only as toys and automata. In fact, *Spiritual Mechanics* begins with a Sufi ritual of mantra-like repetitions and magic words that aimed to protect the work from catastrophe or curse. Taqī al-Dīn later mentions that mechanics reverberates with alchemy and magic because it manipulates the laws of nature with "tricks" and lets man control nature.[28] After dedicating *Spiritual Mechanics* to his patron ʿAlī Pasha, the Ottoman governor of Egypt,[29] Taqī al-Dīn states his aim "to build a machine and a clock that would reflect the spiritual structure of the heavens."[30] Accordingly, in his diagrams of the clock mechanism, he refers to the different wheels as replicating celestial spheres and their locations in relation to the zodiac.

Taqī al-Dīn seemed to think of mechanics in two dimensions: as being useful to and improving life and, more importantly, as revealing the spiritual structure of the heavens. With his connection of mechanics to alchemy and magic, the structures of mechanics and the structures of the world would act in cooperation. Thus, the mechanical unfolding of the structures and secrets (or the secret structures) of the universe could change the cosmic order.

He particularly stressed the connection between mechanics and hermeticism, in arguing that the machine reflects the heavens and that, as a science of the manipulation of the laws of nature, mechanics can also manipulate the influence of the stars and change the cosmic order by incorporating arts of alchemy, magic, and talismans: "[The art of building clocks] relies on two sections of sciences, mathematics and natural philosophy. As for mathematics, it uses fields of algebra, geometry, science of surveying, dynamics, and the science of balances. As for natural philosophy, it requires knowledge in the art of talismans, magic, and alchemy. Both require a high ability of direct intuitive perception, power of imagination."[31] God, the essence of truth, is represented through the perfect motion of the celestial bodies and the resultant creation of time. Thus, the art of clock-making aims to recapture the role of God in the creation. To achieve precise control in timekeeping would be to grasp the essence of nature and allow control over nature.

Such a conception of a mechanical clock that separated the measurement of time from place and nature could bring, under one principle, all creation, from the beginning to the end of time. It also reflected messianic and apocalyptic sentiments, notions that ran in opposition to those of European powers. The poem from *Shāhinshāhnāma* tells us that, for the sultan, "things changed completely" and the wheel of the earth became an immense roving-ground for him. Taqī al-Dīn's role at the observatory, therefore, was not only to help control the "roving-ground" but also to measure the flow of time and to decode the reasons for the rapid changes in world politics. And his knowledge of alchemical methods of cosmic influence was aimed at changing those politics. He expressed the notion of the end of time, or the end of the world, in his major work *The Tree of Ultimate Knowledge in the Kingdom of the Revolving Spheres—The Astronomical Tables of the King of Kings [Murād III]* (*Sidrat al-muntah al-afkār fi malkūt al-falak al-dawār al-zīj al-Shāhinshāhī*) (here, *ultimate* means both the end of time and the end of the world).[32] The title reflects a worldwide apocalyptic sentiment within which, and for which, the observatory functioned.

But, let's take a step back. How did Taqī al-Dīn acquire his skill in constructing mechanical clocks? After all, devices with second hands had been around for only a few decades in Italy. His introduction to *Revolving Planets and Revolving Clocks* implies something about the sources of his knowledge. He tells us that, when young, he used to "study the books of other mathematicians . . . I inspected texts in common use, the Spherica of Theodosius, the Elements of Euclid, the book On Equilibrium of Planes of Archimedes, and the books of arts, which have the precise works and texts on mechanics."[33] Although he could get hold of such classics in their Arabic translations and commentaries, Taqī al-Dīn tells us that, for mechanics, he relied on sources "from other religions," that he "gathered their useful fruits," and that "no one in the Islamic world has come to terms with such knowledge." In a later portion of the book, he explains that knowledge of clock-making had for some time been obtained by rote, and he states that his motive for writing the book was merely to document ideas that might fall into oblivion.[34] Given that Taqī al-Dīn presented himself and his work as pioneering in Islamic mechanics, he clearly received some of his knowledge on mechanics, and especially clock-making, from contemporary non-Islamic sources, about which he did not openly talk.

One clue about his sources comes from a non-Arabic word he used for the mechanical clock: *bānkām*. "*Bānkām* is of Persian origin," he explains, "and in Arabic usage it means 'the origin of a thing or something pure, or the last hour of a night.'"[35] Another use of the word is in the history of Islamic mechanics found in a medieval book ascribed to Archimedes, *The Book of Archimedes on the Construction of Water-Clocks* (*Kitāb Arshimīdis fī'amal al-bānkāmāt*), which quite possibly was a source for Taqī al-Dīn. It consisted of an anonymous description of a monumental water-clock. Donald Hill conjectured that the origins of the book are Greek, with additions by Arabic writers who took it as an important source for late medieval mechanics.[36] Ostensibly, by association, Taqī al-Dīn applied the word *bānkāmāt*, given to medieval water-clocks, to mechanical clocks as well. His training seems to have stemmed from a broad and flexible intellectual field, allowing him to appropriate new and somewhat foreign and stigmatized knowledge.

Contrary to Taqī al-Dīn's view, *bānkām* has neither Arabic nor Persian roots, but may be a corruption of a Latin word. In antiquity, Roman timekeepers divided diurnal time into day and night, each having twelve segments. The last hour before the light, just as Taqī al-Dīn suggested, is *diluculum* ("dawn") in Latin, and one can suggest tentatively that *bānkām*, if it is indeed a corruption of something, may come from that Latin word.[37]

Hours in general, and the hour before dawn in particular, have great significance in mythologies of the apocalypse, especially as a metaphor for the last days before the arrival of the messiah. Cabalists would call it, in Hebrew, *ayelet ha-shahar*, and Muslim millennialists would call it, in Arabic, *ashrāṭ al-sā'ah*—the Signs of the Last Hour given by God to indicate the apocalypse. According to Islamic mythology represented in *Ḥādīth* (traditions of the Prophet and early companions), the first sign of the Last Hour would be the fall of Constantinople, and more signs would occur until the final fall of Rome.[38] While living between the fall of Constantinople (1453) and Süleyman's efforts to fulfill one of the last signs of the Last Hour—the fall of Rome and Christianity[39]—Taqī al-Dīn consciously correlated the signs of apocalypse in Islamic mythology (associated with the metaphor of the Last Hour and the defeat of the Antichrist, *Dajjāl,* at the last hour of the night) with the metaphor *bānkām,* the last hour before dawn. In the very first paragraph of *Revolving Planets and Revolving Clocks,* on the art of clock-making, Taqī al-Dīn follows the same traditions that argued that "God holds the science of the Hour."[40] "O, the one who created motion and rest," he writes, "and from the known [celestial] revolutions gave rise to what is hidden [in nature] and set in these spheres and circles the distilled truths and refined signs . . . and directed the noble intelligence to accept the down-flowing manifestations of truth." To unveil the secrets of nature and the distilled signs of the hours, one should use mechanical clocks that are "the most precise, distilled and refined" tools of the measurement of time and motion that represent God's essence in nature.[41]

Taqī al-Dīn appropriated apocalyptic terminology in his introduction to mechanical clocks because mechanical time corresponded to apocalyptic time. The mechanical clocks presented a new perception of homogeneous time. Sundials were fixed to cyclical, eternal motions, and sand-clocks and water-clocks measured only small segments of time. Mechanical time, by contrast, could be abstract and detached from nature and locality; it could delocalize time and draw reference to a larger concept—cosmic time. Apocalyptic time, reflected in mechanical time, transcends nature and culture. It exists in and of itself as God's time, from the creation to the end of the world.

The application of clocks to astronomy played a significant part in a new perception of time, and Taqī al-Dīn engaged with it not merely for the acquisition of new knowledge. More importantly, it entailed a reform in the methodology of science, or, as he called it, *arṣād jadīda* (new observation): "its subject concerns the special motions of the special bodies that intercept certain distances in certain times."[42] The work on clocks brought about a new perception of nature by breaking down nature into "specificities" and at the same time replacing "worn-out astronomical tables" with new, precise observations. He parted from tradition and inaugurated a "new science," as the poem in *Shāhinshāhnāma* claims: Taqī al-Dīn surpassed the achievements of the most prominent late medieval Islamic astronomers and mathematicians, Ibn-Shāṭir and Jamshīd al-Kāshī. He, arguably, constructed better instruments and, by using compasses and mechanical clocks in his observations and calculations, he came up with precise astronomical tables.[43]

Taqī al-Dīn's universe goes beyond the traditional and parochial perceptions of Islamic observatories. He implied that some of his knowledge came from foreign sources, but there is no explicit textual evidence to support this. Why did

Taqī al-Dīn conceal his awareness of European achievements? Traditional interpretations that hint at some sort of intellectual persecution are extraneous in our case, because we do not know of a clash between heliocentrism and Islamic religious institutes or any religious scientific dogma. The answer instead should be sought in the "trustworthiness" and the credibility of Taqī al-Dīn's channels of circulation of texts and objects.

The Mysterious Jew in Taqī al-Dīn's Toolkit

Returning to the *Shāhinshāhnāma* miniature (see figure 1), we see in the top row two figures who are significantly bigger than the others. One man holds the astrolabe while discussing something with the other. The painter may be emphasizing their seniority at the observatory, and, indeed, the text of the inscription tells us that only five of the fifteen staff members ever acquired sufficient knowledge and skills in the practice of astronomy. Moreover, the emphasis on two members, more or less equally, indicates that Taqī al-Dīn was chief, but that he had a peer; some scholars suggest that this might be Khawājā Sa'ad al-Dīn, the sultan's tutor, from whose house the astronomers observed the eclipse of 1574, while the observatory was still under construction.[44] But, it could be another senior fellow at the observatory.

Taqī al-Dīn's *Tree of Ultimate Knowledge*, which describes the results of his work in the observatory, mentions three eclipses during his time in Constantinople. The last could not be observed from Istanbul, due to clouds, and consequently a Jewish astronomer that Taqī al-Dīn mentioned mysteriously as Dāwūd al-Riyāḍī, "David the Mathematician," supplied the missing data.[45] David conducted his own observations in his home town of Salonika and supplied corroborating data to Taqī al-Dīn.

In the Jewish Sephardic communities, "linguistically skilled" Ottomans reworked Latin texts of astronomy. After expulsion from Spain, many Jews migrated to the eastern parts of the Mediterranean, first to Italy and then to Salonika and Constantinople. By the second half of the sixteenth century, Salonika had become an economic and cultural center for refugee Jews, namely, those who fled the Italian Inquisition in the 1560s—physicians, artisans, and philosophers literate in Latin,[46] such as 'Arama, De Leon, and Amatus Lusitanus.[47]

The enigmatic figure of David the Mathematician has produced some unempirical scholarly speculations.[48] The Jewish community of Salonika is well documented. However, in the lists of books produced by its local presses, we find no mention of this David, so he probably did not publish a book. It is reasonable to assume that a man from Salonika would have left some life traces. Yet, in the list of tombs in the local Jewish cemetery, no David fits the few clues we have.[49] Perhaps he left Salonika later in life and died elsewhere.

In an introduction to a contemporary Hebrew book on the new cosmologies, David's full name surfaces. Jacob HaLevi, an Italian Jew who migrated from Venice to Salonika in the 1560s, clears up the mystery. In a passage of approbation for *Sefer Elim* (1628), a book of astronomy and mathematics written by Galileo's student Joseph Solomon Delmedigo, HaLevi writes about his childhood teacher: "I remember in my youth days, while I was pouring water on the hands of my teacher, the wide-ranging scholar Rabbi David Ben-Shushan, he was asked by the great wise scholars of the Muslims questions about trigonometry, metaphysics, astronomy and philosophy."[50]

In another contemporary account of Jewish scholars of the sixteenth century, written by David Conforte, the sixteenth-century compiler of biographical works on Jewish scholars, David Ben-Shushan is mentioned as "a wide-ranging scholar in all the fields of wisdom . . . astronomy and philosophy as well as in Islam, to the extent that occasionally the great Muslim scholars were referred [to him] for advice. He moved to Constantinople in 1574, and the great Muslim scholars used to honor him there for his great wisdom."[51] In the 1550s, Ben-Shushan fled from Venice to Salonika, where he taught mathematics. His interest in natural philosophy is revealed in another manuscript that was copied in mid-sixteen-century Salonika under the title *Toldot HaAdam* (The Generations of Mankind),[52] in which he translated Thomas Bricot's *Textus abbreviatus Aristotelis super VIII libros physic*, a digest of Aristotle's philosophy. Ben-Shushan also included subtitles from other Aristotelian works such as *De caelo*, *On Generation and Corruption*, *De anima*, and the *Metaphysics*. The translation is from Latin into Hebrew, although the diagrams from *De caelo* also include some Arabic insertions.[53] Moreover, Ben-Shushan's commitment to the premises of *De caelo*, including his belief in the eternal world, can be seen from a single autobiographical insertion at the end of the essay, where he writes: "the book of *De caelo*, from which no secret escaped, is now completed."[54] Evidence suggests that Ben-Shushan, or Dāwūd al-Riyāḍī (David the Mathematician), who moved to Istanbul in 1576–77 and joined the observatory's staff, not only supplied Taqī al-Dīn with the missing data on the eclipse that could not be observed in Istanbul,[55] but also transmitted to him some Latin knowledge on natural philosophy.

Any interest in David Ben-Shushan should not end simply with recovering his identity. Sephardic Jews such as Ben-Shushan, HaLevi, and other Salonikans, in addition to their cross-cultural character, held a cosmic perception that the expulsion from Spain in 1492 signified the coming apocalypse and the imminent return of the Messiah. The intensive cabalistic discussions of that time period even portrayed the Ottoman sultan, Süleyman the Magnificent, as a tool in the hands of God that would bring vengeance on Christendom,[56] and stimulated the printing of some ancient apocalyptic texts.[57] In a sense, the expulsion from Spain and, later, the persecutions by the Italian Inquisition pushed Jews toward the Ottoman Empire—Jews who looked at the sky and at world politics for confirming signs of their redemption. However, connections with Europe continued, and many Jews kept moving between Italy and Salonika for trade, and even to Venice for publication of books. The blend of messianic apocalyptic sentiments and cross-cultural identity embodied the social makeup of the men surrounding Ben-Shushan. In fact, his student HaLevi was held captive by Christian pirates in Malta. After gaining his freedom, he moved to Salonika and became head of an organization that raised money for ransoming Jewish captives held in Italian jails.[58]

In Salonika, David Ben-Shushan also engaged in astronomical activity. He noted observations of eclipses, and his interest in Latin natural philosophy reflected the cultural context in which scholars who translated masterpieces in astronomy from Latin to Hebrew traveled. In the 1560s, for instance, Moshe Almosnino translated from Latin to Hebrew and Ladino (the Jewish Spanish dialect) two medieval and late medieval astronomical manuscripts: *Sphaera mundi*, by Sacrobosco, John of Holywood, who mostly repeated the writings of Muslim astronomers; and *Theoricae novae planetarum*, by George Peurbach, who despite his dependence on Arabic sources, disregarded them and called for a return to the Greek sources of astronomy.[59] Taqī al-Dīn would have been familiar with these translations through his colleague David Ben-Shushan.

What can we learn about the acquaintance between Taqī al-Dīn and David Ben-Shushan? How did Taqī al-Dīn know him, and why would he trust him to impart correct and credible knowledge? Taqī al-Dīn could have known him through a third party. The head of the Jewish community in the Ottoman Empire, Don José el Duque de Naxos (named, in Hebrew, Don Yosef Nasi), was affluent and influential in the Ottoman court.[60] He maintained a tight network of assistants and connections throughout Europe, to the extent that Sultan Selim II (the father of Murād III) used his services in diplomacy and helped him conduct commercial interactions with various courts. Nasi was connected to court politics in Constantinople, to the Jewish community of Salonika, and to various European cities in a network of economic, cultural, and political exchanges. Gifts, books, and visitors (especially between Venice and Urbino) moved back and forth in a network directed by Don Yosef Nasi, exposing its members, such as Ben-Shushan, to European culture and knowledge. David Ben-Shushan, then, was most likely the secret Jew in Taqī al-Dīn's toolkit. However, learning this does not tell us much about Taqī al-Dīn's European sources before he moved to Constantinople—that is, during the 1550s and the 1560s.

Schweigger's Journal: The Master and the Captive

Another network connected Taqī al-Dīn with Europe—pirates and their captives. Salomon Schweigger (1551–1622), Habsburgian envoy to Istanbul, closely witnessed the controversies engendered by the Ottoman Observatory, and in his journal covering the 1580s there are clues about the intellectual background of Taqī al-Dīn. In an unsympathetic tone, Schweigger writes about the way "the Sultan was led into great expenses by a worthless Astronomer."

> When we arrived at Constantinople an Arab convinced Sultan Murād, who was unusually inclined to the good Arts, that if he was granted permission and was to receive all necessary assistance, he would undertake such an endeavor, as would enable him to predict for the Sultan, future events through examination of the stellar constellations. Since, however, such a project requires an enormous effort, it shall be necessary for the Sultan to cover the expenses, which the Sultan quickly agreed to. He also received a daily allowance, [and] rumor had it that twelve captured Christians were also being kept for him to assist him with his work. A dwelling place was built for him in the desert outside the city of Galata, wherein the conjurer should craft his prophecies so that he may pursue unhindered his interest in deception.[61]

We learn that Taqī al-Dīn had "twelve captured Christians" as assistants,[62] but surely he had cultivated his mechanical skills well before this. On possibly earlier sources of knowledge, especially in mathematics, astronomy, and mechanics, Schweigger adds another important piece of information: "As regards his personality, he was an artless charlatan, unholy rogue. Who was kept as a *prisoner in Rome* many years ago by a mathematician whose servant he once was. This is where he mastered his art and became an artist of the heavens and conjurer of constellations. He obtained the writings of Ptolemy, Euclid, Proclus, and other famous astronomers in the Arabic language and *secretly kept a Jew* to explain these writings."[63] Taqī al-Dīn, according to Schweigger, had questionable reliability. Undocumented as they were, captives were excluded from historical records, and thus Schweigger's doubts about credibility cannot be corroborated. However, we have unexpected firsthand evidence for Taqī al-Dīn's Italian connection. In one of his manuscripts, discovered in the National Library of Tunisia, George Saliba found marginalia, in Taqī al-Dīn's handwriting, in which he states that it would be better to consult Italian sources and dictionaries.[64] Thus, Taqī al-Dīn knew Italian and was exposed, somehow, to Italian culture.

European captives in the hands of Muslims and Muslim captives in the hands of Europeans reflect two sides of the same coin, a phenomenon of piracy and war. Conceivably, Taqī al-Dīn fell captive while sailing from Alexandria to Istanbul sometime between 1549 and 1552 or in the 1560s. Salvatore Bono supplies us with ample evidence of ships caught by Italian pirates.[65] A vivid source from the late sixteenth century is the memoirs of Muṣṭafā Efendi, who was captured while sailing from Istanbul to Cyprus and kept records of his two years' captivity in Malta. Malta was an encampment for thousands of Muslim captives who were sent as slaves to Italy. Moreover, the number of years in prison was determined by the rank of the captive, and because Muṣṭafā Efendi was a *qāḍī* (judge), just like Taqī al-Dīn, he was able to raise the ransom more easily and was released after only two years.[66] Kidnap and ransom exchanges became an active phenomenon, and, in fact, Sultan Murād III issued two royal decrees (*firmans*) in 1592 concerning the exchange of captives with Europeans.[67] From Taqī al-Dīn's and even Schweigger's writings, general evidence indicates that European captives in Istanbul worked as servants for scientific projects such as building clocks and cannons and making astronomical observations.[68]

In addition, there was a specific trend in Muslim travel to Europe.[69] Arab captives in Italy were sometimes put to work in local intellectual projects of Arabic translations.[70] The flow of oriental dignitaries and envoys to Rome also stimulated interest in the Arabic language. Especially influential in Rome was the Arab traveler Leo Africanus (al-Ḥasan ibn Muḥammad al-Wazzān), who was captured in 1517 by pirates, sent to Rome, and temporarily converted to Christianity. His works, especially *De totius Africa descriptione* and *De viris quibusdem illustribus apud Arabes*, greatly increased Europe's knowledge of the Muslims of North Africa.[71]

Since the middle of the sixteenth century, growing missionary and economic interests in the Near East had led to the development of Arabic printing presses. Aware of the popularity of printed works in the Islamic world, Italian printers looked to profit from Arabic versions of the scientific classics. The great interest in the Arabic sources of Euclid brought about a newly printed Arabic *Elements*, something in which Taqī al-Dīn was instrumental, according to Schweigger. Such projects required teams of Arabic scholars who had a profound familiarity with scientific literature and could work for several years on a project. We also know of a world map that was printed in Venice in 1560 and delivered to the Islamic world. In its introduction, Ḥajjī Aḥmad, the probable author, tells a story of woe, according to which he was a suffering captive in Italy. We learn, further, that he requested that his Muslim brothers purchase the map so that the income might be used to set him free.[72]

The stories of Leo Africanus, the map of Ḥajjī Aḥmad, and the artisans of Arabic printing in Italy supply circumstantial corroboration for Schweigger's account of Taqī al-Dīn's having been a servant of an Italian mathematician who used Taqī al-Dīn's linguistic skills to look for Arabic sources of Euclid. Italian scientific culture thus fostered an ongoing need for scholars with access to such knowledge and linguistic skills.[73] The identity of Taqī al-Dīn's master is unclear, although Schweigger's argument is certainly plausible.

Renaissance mathematicians differed from their medieval predecessors in their culturally open-minded environment and in the growth of textual resources. The scientific and political developments of the late fifteenth century—the weakening of the authority of the Latin translations of the Arabic sources on Greek natural philosophy, as well as the fall of Constantinople and the transmission of additional Greek sources—were reflected in Italian libraries. With the loss of Byzantium as a buffer between Europe and Islam, and as the heirs of Greek culture, Italian mathematicians became attracted to the Greek sources. Venice and its libraries became the center of Greek studies and held most of the Greek classical manuscripts.[74] Thus, the new humanist libraries in Italy encompassed not only Euclid and Archimedes in Latin, Greek, and Arabic, but also Greek texts of Apollonius and Diophantus, and Proclus's commentary on Euclid, which were the raw materials for Renaissance mathematics. The combination of Euclid, Ptolemy, and Proclus was of interest to Taqī al-Dīn's master and had been a common set of ancients among Renaissance mathematician-scholars since the late fifteenth century.

For Federico Commandino (1509–75) of Urbino, for instance, the true vocation was editing, translating, and commenting on the classics of ancient Greek mathematics. Commandino began his interest in ancient mathematics by translating from Greek to Latin and commenting on various works of Archimedes. During the 1550s he heard complaints about the difficulty of understanding Ptolemy's *Planisphaerium*,[75] a text that showed how circles on the celestial sphere may be stereographically projected onto the plane of the equator. Such difficulty arose in part from the loss of the original Greek text, and Commandino worked to recover it from an extant medieval Latin translation of the Arabic version.[76]

Like his predecessors, Commandino seems to have been galvanized by the poor state of contemporary mathematics, which he could trace back to the Middle Ages. Accordingly, he looked for sources that would help him recover the Greek texts of Archimedes, Ptolemy, and Euclid.[77] But when he found only parts of these texts in Greek, he turned to Arabic sources, either to update Latin translations from Arabic or to translate works that had not yet been translated. To do so, he had an assistant, fluent in Arabic philosophical texts, who facilitated the translation projects. This student, Bernardino Baldi (d. 1617), was a well-rounded scholar in classical languages, including Arabic. Ostensibly, he helped Commandino access the Arabic sources that he, Baldi, had mastered (to the extent that he eventually wrote a pioneer study on the history of Arab mathematicians)—although he apparently employed a captive in his service.[78] Thus, Commandino's great interest in Arabic sources, Baldi's orientalist work, and a possible Arab captive, taken together, confirm Schweigger's suggestion of the latent role of captives in the scientific culture in Italy.

Muslim and European scholars, then, acted on each other as direct sources of knowledge. However, an exchange of knowledge was promoted not only directly but also indirectly and unconsciously when Ottoman and European cultures engaged in a dialectical interplay of ideas, fears, and emotions. Political and cosmological crises led to practices of astronomy that worked against the background of a historical a priori: the "end-times."

Dialectical Stimulus: Astrological Competition in a Euro-Ottoman Space

Early modern astronomy was extensively supported by rulers, not for their love of the arts, but mostly for their need to relieve political and personal anxieties. In times of crisis, rulers invested vast funds to bring notable astronomers to their courts. In special circumstances, they even promoted the building of observatories, an enterprise that required huge funds and long-term scientific efforts, for at least a couple of decades. The Chinese Empire pioneered in establishing observatories that primarily aimed to serve political needs in predicting the coming of eclipses and their astrological implications. The Mongol-Muslim dynasties extended such political culture westward, when the grandson of Gengis Khan, Hulagu Khan (d. 1265), conquered southwest Asia. He invited the greatest astronomer of the time, Naṣīr al-Dīn al-Ṭūsī, to build the first Muslim observatory in Marāgha (northern Iran), which eventually produced the *Ilkhanic Tables*. Almost two centuries later, the remains of the observatory inspired the Timurid ruler and astronomer Ulugh Beg to invite the greatest astronomers and mathematicians of his age to work in a new observatory that he built in Samarqand in 1428. His staff worked for many years and in 1437 compiled updated astronomical tables that became the conclusive tables for Near Eastern and European astronomers until the early seventeenth century. One of the observatory's leading scholars, ʿAlī Qūshjī (d. 1474), left Samarqand after the death of Ulugh Beg and arrived in Istanbul to continue his work, this time in the service of the Ottomans. A century later, a new generation of observatories came into being, one in Istanbul, the other in Hveen and later in Prague. The two observatories dialectically stimulated each other to practice astronomy, to make astrological predictions about the doom of the other, and to produce new astronomical tables.

Tycho Brahe and the Turkish Antichrist

Tycho Brahe started his career in astronomy by observing a lunar eclipse in 1566 and making a horoscope for Sultan Süleyman the Magnificent. Tycho predicted that Sultan Süleyman—who subjected Europe to horror and fear, and whom Martin Luther referred to as the Antichrist—would soon die.[79] Süleyman died a few days before Tycho's predicted time, however. This was a painful lesson for Tycho. Later, his astrological interpretations of the 1572 supernova, the 1577 comet, and the planetary conjunction of 1592 did not explicitly name the Ottomans but predicted radical changes in world politics and religion.[80] The comet of 1577 "augurs an exceptional great mortality among mankind, the like of which has not occurred in many years." The effects of this comet would be felt "more to all those kingdoms which lie toward the west in Europe within Christendom, . . . and regions pertaining thereto will be disturbed more than the Oriental Turkish and Persian subjections."[81] Given the apocalyptic sentiment, his predictions generally referred to the Ottomans, especially because he correlated the fall of empire with the fall of a particular religion and a cosmological war between good and evil.[82] We learn more about this from later sources that connected Tycho's astronomy with the astrological prediction of the fall of the Turks.

In an anonymous astrological pamphlet entitled *Predictions of the Sudden and Total Destruction of the Turkish Empire and Religion of Mahomet: According to the Opinions of the Lord Tycho Brahe . . .*, which was circulated in Europe in the late seventeenth century, Tycho's predictions against the "Turks" played as a cure for a general anxiety.[83] The pamphlet aimed to show "by the many strange Comets which have appeared of late years, not only in sight of the same *Constantinople*, but even of all *Christendom*, as Fore-Runners of the great changes which shall happen, and particularly on the *Turkish* empire, whose grand fatal years is proclaimed by the Heavens to be so near, at hand, as has been long since observed by Learned Men, by the cause of the sun's Eccentricity." The pamphlet gives an eccentric connection to the fate of polities: "as it was described by *Copernicus*, . . . by a motion of a little Circle, having the eccentrick of its Center in its Circumference, *Georginus Johannis Rhetius* [Rheticus], called this circle, *the wheel of Fortune, by its revolutions*, saith he, *the monarchies of the World received their Commencements and Changes* . . . According to Tycho this eccentricity is already notably encreased." Leaning on Tycho's observations and predictions about the decline of the "Turks," the anonymous author sought further precision: "And whereas such a Mighty *empire* that hath shaken the Remotest regions of the Earth, cannot reasonably be supposed to be dissolved into Atoms, in a Moments Space of Time; and the Fatal Year of that Monarchy appearing by the said Learned Authority to be *anno* 1728." With visions of disease, destruction, earthquakes, and wars, the author finally proclaims "a happy day" for Europe.[84] The document fashioned Tycho not only as a European hero facing the "Turkish threat," which stimulated his projects, but also as an observer who popularized contemporary astronomy. Thus, the "Turkish threat," a widespread concern, played as the vehicle through which Tycho was introduced to laymen.

Tycho's interest in the East and in the Ottomans went beyond apocalyptic sentiment per se. Tycho was also driven by the impressions of Danish students in Italy who traveled to the Near East. He corresponded with the celebrated geographer Giovanni Antonio Magini, who fascinated him with the search for textual as well as geographic sources of ancient astronomy. Tycho suggested to Magini a project involving construction of an observatory in Ptolemy's city of Alexandria, to redetermine the latitude; he had earlier sent an observer to the observation site of Copernicus at Frombork, in Poland, for the same reason. Magini used his connections at the Venetian court to get Ottoman support.[85] The government of Venice, which took the initiative, started the preparation by sending an observer to Alexandria.[86] For his part, Tycho tried to promote the project through a link with the Habsburgs that involved Jacob Kurtz, a close advisor to Rudolf II.[87] However, for reasons not made clear, the joint project did not take off. Tycho also attempted to send to Alexandria one of his best students of astronomy, Gellius Sascerides, and later, apparently, he hoped to send his son, Tycho Tygesen Brahe—an idea that never worked out.

Tycho's astronomical-apocalyptic role against the Turks fitted Rudolf II's political-apocalyptic struggle against the Ottoman Empire, especially as astronomy was incorporated. Equipped with his instruments and anti-Ottoman astrology, Tycho traveled in search of a new patron. Already a central player in European politics, he used old connections in the Habsburg court—Hagecius and Jacob Kurtz—to seek the patronage of Rudolf II.

Rudolf needed an observatory to work out a political cosmology as well as to make a counter-display of power.[88] In the early 1580s, Rudolf, the Holy Roman Emperor, had moved the capital from Vienna to Prague, because of Ottoman raids on Vienna and Hungary. By the 1590s, he faced another crisis. The war against the Ottomans resumed, and plagues and bureaucratic disorientation caused him to stop functioning directly as ruler.[89] In the background, popular interest in the fearsome and ungodly qualities of the Ottoman court found expression in various genres of writing, from the most scabrous news sheets to sophisticated refutations of Islam. Nearly all of them propagandized in favor of the Habsburgs and used apocalyptic undertones.[90] The efforts of the pamphleteers coincided with those of the government in their search for a common anti-Ottoman front, whose natural leader was Rudolf II (figure 3).[91] Moreover, Lutherans, such as Tycho, in Rudolf II's court represented the extreme of anti-Ottoman sentiment. Schweigger, for instance, was the first to translate the Qur'ān into German.[92] The translation inspired a colleague from his Istanbul days, Václav Budovec, to write an anti-Ottoman account describing the Ottomans as the embodiment of the Antichrist.[93] The two Lutheran envoys, Schweigger and Budovec, advocated the Protestant cause of strong support for the Habsburg interest against the Ottomans.

Ultimately, we are building here a context for Tycho's invitation to Prague by exploring the cultural and emotional content of the larger dialectical context that surrounded him. In the continuous struggles with the Ottomans, Rudolf II reinvented himself not only as the savior of Europe but also as the true heir of Constantinople (figure 4).[94] Not just politically, but also culturally, Prague was considered heir to Constantinople. Rudolf II needed Tycho and other men of art and science, not only to predict the "end-time" and the coming "doom of the Turks," but to defeat the Turks culturally (figure 5). Having Tycho in Prague was a display of power in the face of the Ottoman threat. We know that Rudolf II was aware and informed of the Ottoman Observatory through his envoy Schweigger and through *Shāhinshāhnāma*. Thus, the political and religious contest with the Ottomans was also astronomical and cosmological.

Finally, Tycho embedded his works—his astronomical practices, court politics, and astrological predictions—in apocalyptic sentiments of "end-times." After the death of Süleyman the Magnificent (1566) and the Ottomans' first defeat, at

Figure 3. Propaganda during Rudolf II's reign included representations of his triumph over the Ottomans. In this picture, distributed as a pamphlet, the eagle signifies Rudolf II defeating the "Turkish" sultan. The pamphlet was printed in 1598 in the old city of Prague by George Datscheikfn. For other examples, see Karl Vocelka, *Die politische Propaganda Kaiser Rudolfs II: 1576–1612* (Vienna: Verlag der Österreichischen Akademie der Wissenschaften, 1981).

Lepanto (1571), the apocalyptic sentiments among the Ottomans became similar to those in Europe.

The Quest for an Astrologer-Mechanic to Save Murād III's Reign

Taqī al-Dīn was nourished intellectually both by scientific objects that circulated in the Near East through gift exchanges with European courts and by Italian intellectual sources. Moreover, cross-cultural exchange was two-way. Tycho did not

Figure 4. Bellona, by Jun Müller (1600). The all-Christian role of Rudolf II, in front of the Ottoman, is represented by Bellona, the goddess of war, who leads the Habsburg standard into battle against the Ottomans. The plate is dedicated to Rudolf II's brother, Matthias, as leader of the Christian armies. As Bellona stirs up battle with her bronze trumpet, Mercury (*Caducifer*) exhorts the sons of Romulus to lay low the Muslim forces. Courtesy of the Trustees of the British Museum.

Figure 5. The Three Goddesses of Arts (1597), by Jun Müller. The connection between the "Turkish threat" and the rise of arts and sciences in Rudolf II's court is represented in this print. The "Turkish threat" to western Europe was particularly serious in the 1590s. In 1596, the great battle at Mazö-Keresztes had resulted in a victory for the Ottoman forces. The three arts—Architecture, Painting, and Sculpture—are attended here by princes and prelates (bottom left). In the background, a galley is attacked by Turkish boats; the Ottomans advance (bottom right). Beset by the Ottoman menace, the three sister arts (*Sorores*) ascend, blown by the flower-scented Zephyr and accompanied by trumpet-blowing Fame, to Jupiter, who symbolizes the refuge offered to artists by the Parnassus created by Rudolf II in Prague. There they will remain, until, as the inscription makes clear, "obtineant sua pristine regna"—they can regain their former kingdom. Courtesy of the Trustees of the British Museum.

completely overlook the Near East: his Muslim reference points were not merely highly desired Arabic texts and the drive to corroborate observations made in Alexandria. Competing apocalyptic visions stimulated his astronomical project. But what were the motives of his adversary Taqī al-Dīn and for the establishment of the Ottoman Observatory?

Establishing an observatory entailed huge investments, and the sources indicate that the Ottoman Observatory received these resources from the court. The motivations behind court funding for such institutions vary from one culture to another. In China, for example, observatories tended to be ongoing efforts to corroborate the ruling dynasty's ability to display its superior knowledge of nature, as well as to gain political legitimacy by employing precision in court rites. Tycho's observatories, first in Hveen, Denmark, and later in Prague, were a display of power; but power in his case was expressed through a coalescence of individual scholarship and careers with a pan-European religious sentiment that required the wealth and power of courts to carry out its mission.

In the Islamic world, the two earlier observatories—the one at Marāgha in the thirteenth century and the other at Samarqand in the fifteenth—were embedded in the dominant Mongol-Chinese political culture, which demanded of the bureaucracy a high degree of precision in its predictions and wisdom in its interpretations of astral events. We do not know of a need for, or an application of, long-term astronomical observations in the Islamic world before its encounter with the Mongols. The setting up of Taqī al-Dīn's observatory, however, was not in that older, Mongol-Chinese mold. It was not part of a long institutional tradition, but rather, to use Schweigger's words, an Ottoman bureaucratic "adventure." Motives may be found in the urgent political needs that appeared in 1574, the year of Murād III's accession, as well as the year when Taqī al-Dīn was invited to Constantinople. Apocalyptic sentiments and the tradition of hermeticism were stirred up by celestial and calendric events—comets, eclipses, planetary conjunctions, and the forthcoming Islamic millennium.

Cornell Fleischer notes that Mevlānā 'Isā, a hermetic and millennialist Ottoman scholar with strong connections to court circles under Süleyman the Magnificent, wrote three recensions (in verse) of an Ottoman history called *The Compendium of Hidden Things* (*Cami-ül Meknunat*). Mevlānā 'Isā's apocalyptic vision relied on the forthcoming 1564 conjunction of Jupiter and Saturn, which signified the end-time. Accordingly, after the conjunction, Süleyman the *Mujadid* ('Isā used the Islamic traditional name for persons who made the transition from one epoch to another) would renew Islamic religion and bring it to its final messianic stage—a spatial universalization to be brought about by worldwide

Islamic rule. The source for the astrological perceptions of the conjunction, as for Tycho and most contemporary astrology, was Abū-Maʿshar, whose work *The Book of Religions and Dynasties* (*Kitāb al-milal wa al-duwal*) discussed how a planetary conjunction indicates the decline and replacement of a specific religion or empire and represents the end of one historical cycle and the start of another: "if the conjunction and its lord are in cardines [i.e., cardinal points] of the malefics [planets that symbolize damage and loss, mostly Saturn and Mars] or in cardines of retrograding and cadent planets, this indicates the shortness of their period, the extinction of their dynasty, and their death."[95] According to Abū-Ma'shar, such events as the Exodus, the birth of Jesus, and the rise of Islam occurred during conjunctions of Jupiter and Saturn.

Astronomers, astrologers, and rulers in both the Islamic world and Europe intensively read Abū-Maʿshar's work. Moreover, contemporary messianic astronomers in Europe mentioned the planetary conjunction as a counter-explanation of the death of the Antichrist. Predictions were tested in 1564, when a series of conjunctions among Mercury, Venus, Mars, Jupiter, and Saturn appeared in the sky above European and Ottoman cultures. The astrological struggle was mounted: Süleyman was either the Messiah or the Antichrist; either he would gain final victory over Europe, or Europeans could expect Süleyman to disappear from history. Given that Süleyman died two years later, without fulfilling the messianic promise, the sky seemed to favor Europeans.

Although Süleyman continued to lead the Ottoman armies until his death during a campaign against Szigetvar-Hungary, his messianic ideology and the general perception of him as the *Mujadid* became blurred. With the *Mujadid*'s death, his successors had to decide whether to break with or to carry on his legacy. Because Süleyman had eliminated two of his promising, competent sons, only his incompetent, unnoticed son, Selim, was left. The latter, crowned in 1566, had not been trained in the workings of the bureaucracy or of the army and court. Faced with an unbridgeable gap between himself and his late father, Selim did not try to carry on the messianic ideology. He was the first sultan to withdraw from the everyday running of the empire, leaving it to his Grand Vizier, Mehmet Pasha Sokullu. The radical shift from Süleyman the Magnificent to "Selim the Drunkard," as contemporaries derogatorily called him, marked the "signs of the hours" and the beginning of the decline of the Ottoman Empire.[96]

The changing cosmic order also had its signs on the battlefield. In 1571, only four years after the coronation of Selim II, the Ottomans suffered their first defeat. At the Gulf of Lepanto, a European coalition fleet including Venetians and Spaniards destroyed the Ottoman fleet commanded by ʿAli Pasha (figure 6). The

Figure 6. Battle of Lepanto (ca. 1572), by Paolo Veronese. In that battle, the Turkish fleet was defeated, thanks mainly to the Venetian ships. The play of tone and light in the lower part, depicting the battle, is masterly. In the top part, above a curtain of cloud, the saints Peter, Roch, Justine, and Mark implore the Virgin to grant victory to the Christian fleet. In answer to this, an angel hurls burning arrows at the Turkish vessels. Courtesy of Gallerie dell'Accademia, Venice.

Ottomans lost more than three hundred galleys, and fifteen thousand men were either killed or captured. Moreover, the European victory led to the release of almost a thousand Christian captives held as ship oarsmen. The battle of Lepanto was presented in contemporary European accounts as a sign of the defeat of the Antichrist, and in Ottoman sources as a catastrophe that marked the forthcoming end of time.[97] The catastrophic eight years of Selim II's rule corresponded to an acknowledged cosmic change. In addition to the question of the planetary conjunction, Mevlānā ʿIsā presented a millennial-apocalyptic account that would support the claim that Süleyman was the watershed beyond which lay the Last Days. Thus, Süleyman, the tenth sultan, who ruled during the tenth Islamic century, was supposed to inaugurate the millennium in the year 1000 of the Hijra (1591 / 1592): the result would be a world under Islamic justice, as well as other extraordinary events.[98] However, Süleyman did not make it to the 1590s; he died in 1566, and his successors, especially Murād III, whose reign (1574–95) reached the millennium, somehow had to cope with the predicted events. Murād III could not shirk from the prevailing messianic ideology that was bolstered by Mevlānā ʿIsā's close connections in the Ottoman court and by the general apocalyptic sentiment infecting the Euro-Ottoman space.

According to the sixteenth-century Ottoman court historian Muṣṭafā ʿAlī, in 1574 there was "disorder of the age and perturbations of space and time which appeared, one by one, after this ruler's [Murād III] accession, and which proved to be the cause of the disruption and degeneration of the order of most of the world."[99] Murād III's cultural taste was different from that of his ancestors. Instead of leaning toward an intellectual literature in the intellectual languages of that time, he favored translations of Arabic works into Ottoman-Turkish, especially works on popular esoterica.[100] Moreover, European sources described Murād III as comparatively weak and superstitious.[101]

Muṣṭafā ʿAlī attached a cosmic-apocalyptic sense to his low evaluation of Murād III's reign. His historiography expressed the general anxieties over the deteriorating condition of the empire as a reflection of the apocalyptic moment. In one of his poems, he depicted the growing corruption of the bureaucracy as the demarcation line between the classical era and the existential political crisis in Murād III's time.

> what a beautiful age was that fine era,
> [for] the clean and the dirty were clear to people
> now we have come to a time
> when neither the incapable nor the noble is distinct.

no one rewards the people of dignity
rather, they are mocked and betrayed.[102]

In his introduction to *Essence of History*, Muṣṭafā ʿAlī portrays Murād III's reign in the grim colors of the approaching apocalypse. The sultan's greatest fault was his greed, which ʿAlī suggested was reinforced by a desire to store riches against the predicted upheavals. ʿAlī portrays Murād III as living between two poles: he was inclined toward the ascetic and unworldly ethic of the Sufi dervish, but he also loved wealth. ʿAlī mentions the sultan's irresponsibility and susceptibility to pernicious influences; the destruction of authority and respect for government; the decline of learning; the sultan's inability to distinguish between true and false spirituality; the corruption of finances since the death of Don Yosef Nasi (1583); inflation and economic crisis; and military disorder.[103] By and large, his historical account was framed by an apocalyptic perception of time: Murād III's reign was leading deterministically toward the end. To make matters worse, certain European writings argued that the Muslims knew of their coming destruction from contemporary European prophecies. Some of the latter prophecies even foretold the breakdown of the Ottoman Empire, beginning in the Islamic millennium.[104]

Standard associations between eclipses and the instantaneous deaths of rulers troubled Murād III.[105] His predecessors Süleyman and Selim II had died a few months after lunar and solar eclipses were observed from Constantinople. Tycho made his prediction of Süleyman's death based on the lunar eclipse of October 1566. On May 20, 1574, six months before the death of Sultan Selim II, a solar eclipse with the longest duration of the sixteenth century was observed in parts of Europe and Constantinople. In addition, the 1570s contained an unusual number of celestial phenomena—a new star (1572) and a comet (1577). All contributed to Murād III's general passion for reading the sky and particular obsession with eclipses (figure 7).

Out of this stellar restlessness came the invitation to Taqī al-Dīn to establish an observatory. He already had spent time in Italian captivity. He fashioned himself as a hermetic mechanic, a master of automata and talismans. Newly appointed as court astronomer, he took charge of fashioning instruments and clocks; updated and corrected the latest Muslim astronomical table of the sixteenth century, Ulugh Beg's *Zīj al-Sultani*; and named the corrections for his patron, Murād III: *The Tables of the King of Kings* (*al-zīj al-Shāhinshāhī*) (figure 8).

As we saw earlier, ʿAlāʾ al-Dīn al-Manṣūr's poem in *Shāhinshāhnāma* supplies ample evidence for astrological motives in establishing the observatory. But more specifically, with "the end of time" at stake, the primary purpose was to update

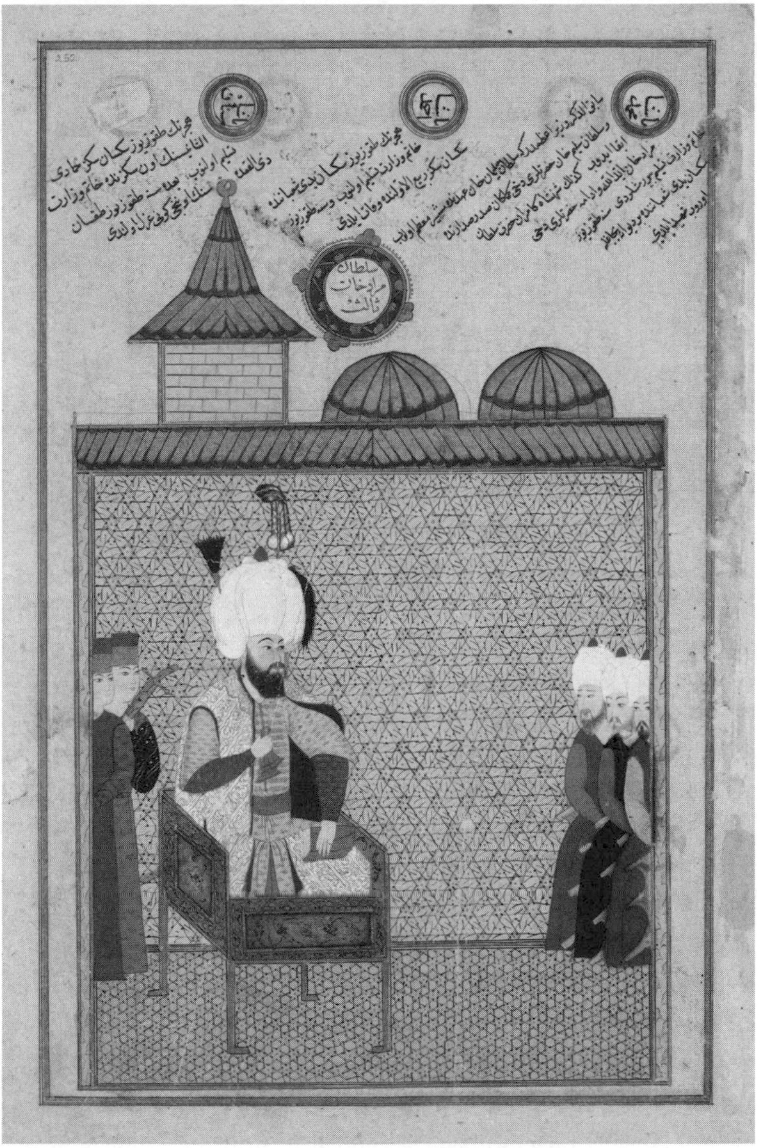

Figure 7. Murād III commissioned several great albums filled with miniatures. Besides the chronicles *Shāhinshāhnāma*, these included an astrological album and *Zübdet al-Tavarih*, in which the great events of his reign were narrated along with astrological events. This miniature depicts star-like prophetic verses hanging from the sky above Sultan Murād III, heralding the rise and fall of the rules of the previous sultans. From *Zübdet al-Tavarih* (Istanbul, 1580), p. 253r, Chester Beatty Library, Dublin.

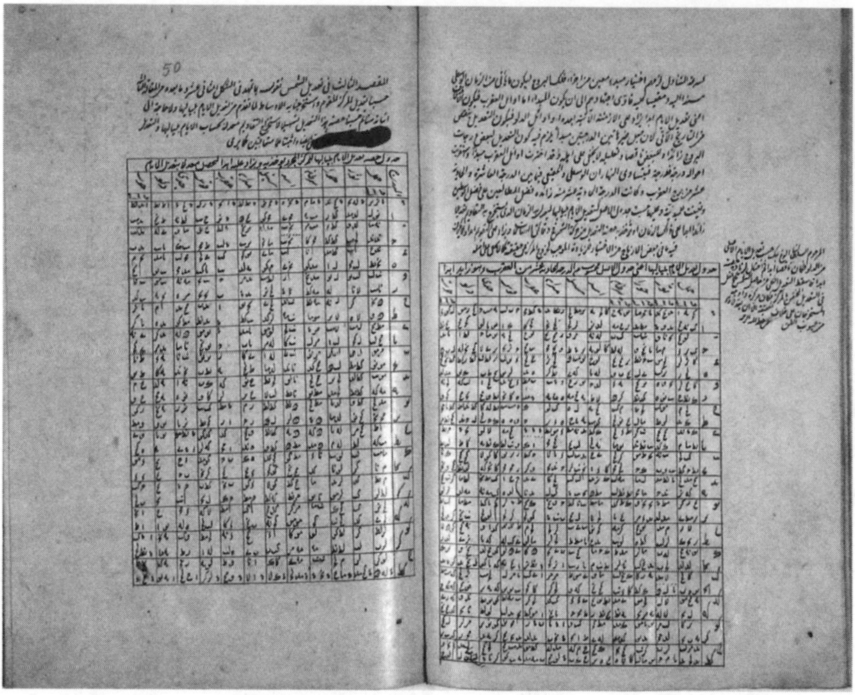

Figure 8. Corrections and updates of Ulugh Beg's astronomical tables, which Taqī al-Dīn named after Sultan Murād III, *al-zīj al-Shāhinshāhī* (The Tables of the King of Kings). Sultan Murād III's fear of eclipses was embodied in the activities in which the Ottoman Observatory was engaged. Apart from updating and correcting past astronomical tables, the manuscript also discusses three eclipses that occurred between 1574 and 1580. From *Sidrat al-muntah al-afkār fi malkūt al-falak al-dawār al-zīj al-Shāhinshāhī*, MS 2930, Nuruosmaniye Library. Courtesy of Süleymaniye Kütüphanesi.

the astronomical tables. The poem mentions that "whoever compiles the new tables during his [Murād III's] reign will become freed from observation programs till the doomsday."[106] Sultan Murād III made his move: "In his delightful era and pleasant age, on a fortunate day and at an auspicious time of his sublime threshold, to his imperial capital . . . came from Cairo a qāḍī [Taqī al-Dīn] of high merits."[107]

The end-time occupied historical writings of the sixteenth and seventeenth centuries.[108] Taqī al-Dīn previously had lived in Egypt, where apocalyptic sentiments were strongly embedded in intellectual culture. Egyptian chronicles saw history through the end-time. For example, Aḥmad Shalabī Ibn ʿAbd al-Ghanī's (d. 1737) eighteenth-century *Clarification of the Signs* (*Awḍaḥ al-ishārāt*) mentions that the science of historiography was given to scholars by the prophet Daniel, who was strongly affiliated with the apocalyptic prophecies, as a way to measure the signs of events made by the seventh sky.[109] Moreover, the eighteenth-century chronicler Murtaḍá al-Zabīdī devoted a long entry to "Constantinople" in his account of Ottoman history, mentioning the fall of Constantinople as the first sign of the "signs of the hour" (*ashrāṭ al-sāʿah*). Later signs, such as the fall of Crete and other European lands, all indicate the fall of Rome and Europe in general.[110] Sambari, a Jewish-Egyptian chronicler of the seventeenth century, presented apocalyptic sentiment by portraying the Ottoman conquest (1517) of the Holy Land and Egypt by Sultan Selim I as one of the signs of the "end of time."[111]

As we learn from ʿAlāʾ al-Dīn al-Manṣūr's poems, Taqī al-Dīn was brought to Istanbul for his unique polymathy, and not merely for his skills in mathematics and astronomy. In Istanbul, the intellectual culture greatly appreciated the presence of numerous students of the mathematician Khawājā Saʿad al-Dīn, the sultan's tutor, next to which "Pythagoras [would be] ashamed of his shortcomings . . . and Archimedes has inevitably gone into hiding . . . and Hipparchus constitutes the type of work done by the least of our Khawājā's disciples."[112] Taqī al-Dīn was exceptional in combining astronomy, astrology, mechanics, and hermeticism, to the extent that he potentially could manipulate the cosmic order by "untangling the knots from Pleiades," as the poem puts it.[113]

Taqī al-Dīn worked not only to unfold the end-time but also to change its outcome. In *Shāhinshāhnāma*, ʿAlāʾ al-Dīn al-Manṣūr writes that Taqī al-Dīn "caused the seven planets and the countless fixed stars to embody and reveal boundless wisdom."[114] However, beyond the tediousness of updating the astronomical tables in preparation for the Islamic millennium, the extraordinary comet of 1577 called Taqī al-Dīn's astrological skills into service (figure 9). The comet, appearing in the year 985 of Hijra, seemed to foretell the fate of Murād III, because its shape was "like a Turban sash over the Ursa Minor stars."[115] Different cultures variously interpreted the comet's shape, its procession along the zodiac, as well as the direction of the tail. These simultaneous observations of the comet informed the political competition between the Europeans and the Ottomans. The mystical nature attributed to comets was used to analogize good and evil. In addition to the comet appearing like the sultan's turban, its location in the sky gave an additional indication of its importance for the Ottomans. As Abū-Maʿshar suggested, when a comet parallels the northern node, it indicates disasters, riots, deaths of kings, "together with a disaster hitting the Turks."[116]

The depiction of the comet's shape and location in *Shāhinshāhnāma* was no different and showed the comet's pertinence to Ottoman relations with neighboring

Figure 9. An astronomer observes the comet of 1577. From ʿAli Muṣṭafā, Nusratnāme, Hazine, MS 1365, p. 5v. Courtesy of Topkapi Palace Museum Library.

cultures. We read that the comet "sent a gush of light from the east to the west, and its appearance was in the house of Sagittarius, its arrow promptly fell upon the enemies of the Religion. As its tail extended in the direction of the east, it discharged its inauspiciousness like a scorpion upon the enemies." Yet, according to the poem, Taqī al-Dīn's interpretation of the comet had to relieve apocalyptic anxieties: "Oh world-swaying king! The candle of your pleasant society shall be resplendent."[117]

Taqī al-Dīn put his prestige at risk in adding an active prediction: "there are joyful tidings for you concerning the conquest of Persia."[118] Such advice concerning war against Persia showed that he worked not only to foretell events and reveal the cosmic order in relation to Murād III's empire, but also to change things. The beginning of the poem explicitly makes the connection between astronomical practice and changing the cosmic order, mentioning the role of the observatory and the astronomers "to carry off the crown which is the comet in the shape of the Sultan's turban from the stars of Ursa Minor."[119] The comet was observed from Istanbul, with Ursa Minor as its backdrop. Taqī al-Dīn's writings on mechanics mention it as a field that represents the divine structure of the cosmos and at the same time brings control over the laws of nature. Thus, the science of mechanics was used in Taqī al-Dīn's observatory not only to reflect the divine order, but also, as 'Alā' al-Dīn al-Manṣūr suggested, "to carry the crown of the sultan [Murād] from Ursa minor" and to change the bad luck.

Unlike his father, Murād III played an active part in bringing about a change in the cosmological predicament, as is seen in his personal title. Usually, sultans bore the title *ẓill Allāh*, "the shadow of God," as a standard form of address. However, as Fleischer showed, with the rise of apocalyptic sentiments, Selim I and then Süleyman used other titles, such as *Ṣaḥib Kirān*, "Master of the Conjunction," indicating their exceptional role in the forthcoming cosmological events. Just as Süleyman appropriated new titles that signified his messianic self-perception, Murād III used a nontraditional title, *Shāhinshāh*, "King of Kings."

Murād III also played out his notions on the broader stage of art and popular culture. In 1582, two years after the death of two of his children, he wrote melancholic poems under the name Murādī. In a depressed mood and with a belief in unlucky destiny, in the same year he issued an order to destroy the Ottoman Observatory and held one of the most splendid festivals in court history for the circumcision of his son and heir. Just like Süleyman the Magnificent, who held a festival in 1530 after the failed siege of Vienna to distract people's attention from the defeat, Murād III wanted to obscure misfortunes by a mighty show of strength and authority. The festival featured art, ethnic group rallies, and technological spectacles. One display used a wagon that moved without horses or any other visible means of locomotion.[120] The device was received with some amazement among the spectators; it used either a carefully concealed operator or some kind of clockwork. Mythological creatures, a giant, a dragon, and a sea monster, also played a part. A pantomime group presented a quasi-dramatic performance with the following plot, as described by a European witness: "there was a band of three lutes, a cornet and a violin played by Italian slaves. There was also a man dressed in black, the astrologer, who carried a sphere or ball in his hands."[121]

Thus, Taqī al-Dīn did not work in an intellectual or theological void. He not only had to produce precise observations of eclipses and comets and make astrological predictions, but also had to demonstrate the manipulation of nature through hermeticism and mechanics. He had to frame his results as credible. However, others would express suspicions about the credibility of the hermetic content of his work and the European sources of his knowledge. The dismantling

Figure 10. Astronomers in action at the Ottoman Observatory. The main building of the observatory had a huge armillary sphere, supported on a wooden frame. The man in the middle is using a plumb-bob to adjust the position of the meridian (north-south) circle, which we see almost edge-on. From ʿAlā al-Dīn al-Manṣūr, *Shāhinshāhnāma* (ca. 1581), F 1404, p. 56b, Istanbul University Library.

of the observatory was equally imbued with apocalyptic foreboding, and, even more important, although Taqī al-Dīn and his staff registered numbers (figure 10) and unfolded the workings of the cosmos, they did not fend off bad luck for the court. Murād III suffered personal and military losses. Subsequently, keen observers perceived the observatory, with its hermeticism and new sciences, as having escalated, rather than diminished, the disasters. According to 'Alā' al-Dīn al-Manṣūr, after a disease epidemic, the clergy openly turned against the observatory and convinced Murād III to destroy it. The act of destruction was likened to the apocalypse: "Nothing remained of the observatory but name and memory; and verily, the fate of the world itself shall be a similar one!"[122] Moreover, 'Alā' al-Dīn al-Manṣūr objected to Taqī al-Dīn's exertions of earthly control over the cosmic order: "If you come to possess the Hermetic wisdom, your thread of intellect will declare itself insufficient to cope with the mystery of life."[123] At the end of poem in the *Shāhinshāhnāma*, we find a lesson on the danger of unconstrained possession of knowledge and an explanation of why, for all the philosophers since Plato, knowledge was limited—"a universal argument that one should acknowledge."

We have traced the evidence on overlapping scientific cultures, of the European and Islamic worlds, in the late sixteenth century. During the mid sixteenth century, Sultan Süleyman the Magnificent projected a grand apocalyptic image—as the Messiah for the Muslims and the Antichrist for the Christians. The apocalyptic visions of each side proposed the other's doom. Rulers, theologians, and natural philosophers saw the exceptional phenomena in the sky—a new star, a comet, and a great planetary conjunction—as positive indications. Within an apocalyptic contest, Tycho Brahe became known among Europeans as the greatest of astronomers, whose precise observations could validate apocalyptic prophecies. Accordingly, he predicted the death of Süleyman, the destruction of the Antichrist, the fall of a religion and an empire, and the rise of Europe. On the other side, the Ottomans, who were aware of these prophecies, faced a cosmic crisis after the death of Süleyman and looked to the sky to find indications of the changing cosmic order and to shift the sky to their side. Consequently, as Murād III came to power in 1574, he invited Taqī al-Dīn to build an observatory in Constantinople. Taqī al-Dīn moved from Egypt to Istanbul after he had made his name as a hermetic philosopher and mechanic, a builder of automata, mechanical clocks, and talismans. Probably while in captivity as a savant under service to an Italian mathematician, he acquired additional knowledge of astronomy, mathematics, and the art of clock-making. This, in turn, helped him inflate his own image—enough to become court astronomer. Thus, drawing from Islamic and European sources in the art of mechanics, mathematics, and astronomy, acquired through several sources—education in Islamic *madrasahs*, captivity in Rome as a scholar-servant, friendship with his Salonikan-Jewish colleague David Ben-Shushan, and the receipt of scientific objects through gift exchange—Taqī al-Dīn built instruments, clocks, and armillary spheres, registered observations, and compiled astronomical tables. He used these in his observation of the comet of 1577 to foretell the future and allay the political confusion of the Ottoman Empire. However, the escalation of the sultan's "bad luck" discredited Taqī al-Dīn's practices and led to the dismantling of the observatory in 1581.

These surprising findings make up a nuanced picture of overlapping scientific cultures in the late sixteenth century. The presentation of Tycho as a pure "European manifestation" and of Taqī al-Dīn as an "authentic" Islamic natural philosopher appears inadequate. Instead of the "two linear paths of development" approach, the findings presented here reveal scientific projects that developed dialectically. Animosity between the two cultures brought with it the sense, in many minds, of otherness and oppositeness "over there." But the argument for a complete separation of the two worlds cannot be supported. Military clashes and their aftermaths produced exchanges of captives, refugees, exchanges of scientific objects and books, and diplomatic envoys moving back and forth, carrying intellectual assets from one culture to the other.

Apart from this direct circulation of men and objects, a shared mythology dialectically constructed the two scientific cultures—each seeing the other's doom in the end-time. Simultaneous observations of the same celestial objects and events—the new star, comet, and conjunctions—connected the two cultures. The people of Constantinople, Prague, and Copenhagen shared the same sky and the same astronomical events. They shared a common cosmological space and apocalyptic sentiments that made their visions reflections of each other. As a result, astronomical practices were a tool in the hands of competing apocalyptic visionaries. Competition occurred not only on political and military grounds, but also on cosmological grounds. Each side encouraged astronomy to find the appropriate signs.

Within the "space race" of the late sixteenth century, a wish for the destruction of the antichrist "Turks" stimulated Tycho's project, and on the other side, anxieties concerning signs that appeared to favor the Europeans influenced Taqī al-Dīn's project. Both sides struggled for superiority through observations that would enable the unfolding of the cosmic order.

Exchanging Heliocentrism for Ur-Text

IN 1623, AN ITALIAN TRAVELER, Pietro della Valle, reached the Portuguese colony of Goa, in western India, after nine years of travel in the Near East. In the same year, a Jesuit, Christopher Borrus (or Borri), on his way back to Italy, also stopped in Goa after his missionary work in Cochin-China (southern Vietnam). Della Valle and Borrus stayed in the same monastery and met for the first time at a midday meal. They exchanged views about the various Eastern cultures they had explored. Borrus bragged of how he had impressed the Chinese literati by making accurate astronomical predictions, thus convincing them to convert to Christianity. In response, della Valle mentioned meeting a brilliant Persian astronomer, Mullah Zayyn al-Dīn al-Lārī, who had firmly rejected the possibility of conversion. Borrus then offered to use the same approach that had proved successful in China: to send a translation of his book on the Tychonic system to al-Lārī, with the hope of convincing him that the advanced state of European astronomy resulted from religious superiority. Quickly agreeing, the two men—della Valle, trained in classical and Near Eastern languages, and Borrus, skilled in astronomy, cartography, and mathematics—worked to translate into Persian a short Latin work by Borrus on the Tychonic system. The translation, as it has come down to us, is in the form of a Persian-Italian manuscript letter, made up as a booklet. It was addressed to a Persian astronomer, al-Lārī, and eventually became part of the collections of the Vatican Library. The outcome of this meeting seems to represent an interesting attempt to convert a Muslim by using arguments about the superiority of European astronomy, but the details of the project, as found in the margins of the letter, indicate something rather surprising.

Della Valle's handwriting in the manuscript letter to al-Lārī appears in a column of Italian and a column of poor Persian, but also includes phrases and terms in Arabic, Ottoman-Turkish, and Latin. The heart of the translation sets out the technicalities of the Tychonic system, but certain autobiographical insertions on the margins of the introduction and the concluding sections introduce the possibility that the Copernican cosmology, based on the Galilean discoveries, might be a better world-system.[1]

There is another source for students of della Valle's letter. After returning to Italy in the 1650s, della Valle published a popular travel journal, *De viaggi*.[2] The journal is a collection of letters taken from correspondence with his patron, Mario Schipano. Subjected to the scrutiny of inquisitorial censorship and intended for a large audience, *De viaggi* silences the incident of the letter and accentuates a culturally adventurous della Valle. The journal serves as a backdrop to the manuscript letter to al-Lārī, allowing us to detect the places where della Valle self-censored *De viaggi*, omitting the names of Galileo and Johannes Kepler, whom he mentioned in the manuscript letter.

Finally, in 1662, ten years after della Valle's death, the antiquarian Giovanni Pietro Bellori published a short biography of him that provides a useful third source. Bellori, who later became famous for composing biographies of artists, depicted della Valle as the descendant of a distinguished Roman family, a man linked with leading scholars in Naples.[3] The Neapolitans fostered in della Valle a passion for antiquity that eventually led to his long and famous trip to the Near East. Bellori also supplied some important details that the letter to al-Lārī and *De viaggi* could not include—namely, della Valle's life after returning to Rome and his affiliations and activities among various radical and subversive intellectual societies.

Each of these three main sources is incomplete and provides insufficient clues to our Pythagorean messenger and his motives. The bilingual letter says nothing about della Valle's intellectual background. *De viaggi*, published in Rome for the general public as a curiosity, gives a descriptive account of the trip but is quiet on the question of cosmology. Finally, Bellori's *Vita* addresses della Valle's life before and after his return, but gives us only a glimpse into his social and cultural affiliations. And yet the faint echoes of Galilean discoveries apparent in the letter written in Goa in 1623 provide evidence of a more interesting story of patrons, secret societies, and the subversion of intellectual agendas.

Getting at that hidden narrative brings together two trends in scholarship that have rarely been presented as intersecting: on the one hand, the historiography of the Galileo affair, which touches on European cultural matters relating to science and religion,[4] as well as the inter-élite struggle for patronage,[5] and, on the other hand, the literature of travel and "pilgrimage" to the Near East, often presented as an essentially romantic and nostalgic European flow that had little to do with the intellectual needs of contemporary debates in Europe.[6] Della Valle's story bridges these discursive lines. He traveled in the Near East to search for evidence that could confirm the truth of a single cosmological system—either the heliocentric Copernican system or the combined geo-heliocentric Tychonic system. In his travels, he collected ancient manuscripts of the Holy Scriptures, medicine, and natural philosophy for Schipano and for Schipano's friends in Naples. More specifically, he looked for a Chaldean ur-text of the Book of Job. The search for such a book lay on the fringes of a larger discussion regarding the Bible and Copernican cosmology, with links to contemporary Neapolitan scholars associated with Galileo who claimed that the Copernican cosmology did not contradict the Scriptures and that lost Holy Scriptures actually mentioned heliocentrism. Thus, both Schipano and della Valle saw the ur-text of the Book of Job—which was traditionally thought to have been written very early in history, soon after the biblical Flood—as a historical a priori, an arbiter of disputes between the Copernican and the Tychonic systems. The search extended European controversies across cultures and added pristine history as yet another judge of natural philosophy, along with experiment, mathematics, and Scripture.

Borrus and al-Lārī

From the Persian part of della Valle's letter, we can derive clues about the nature of his relationship with the recipient, the Persian astronomer Zayyn al-Dīn al-Lārī. Instead of opening with a universally appropriate, or generic, greeting, as one might expect of correspondence between two scholars from contesting religions and cultures, the letter begins: "In the name of the son, the father, and the holy spirit the one God," a greeting indicating that della Valle perceived al-Lārī as a candidate for conversion. He then pays respect to al-Lārī's scientific persona, continuing: "To the pride of the wise men and scholars of Persia, Mullana Zayyn al-dīn al-Lārī the Astronomer."

Della Valle proceeds to present the translated work as "an account of the essay of Christopher Borrus, the Christian, on the system of the new world according to Tycho Brahe and other ancient astronomers, written in abbreviation by the humble me, Pietro della Valle, who is famous in Great Rome, and who scribes it from the language of Latin to Persian."[7] He then introduces the Jesuit Borrus and his work.

> [Borrus is] prudent and very educated, and places great efforts in the field of mathematics. He is from Milan, Italy. He taught in general schools and then in order to serve his country was sent to China and in a city close to China,

which we call Cochin-China, where he stayed for a few years, and then went across the ocean on a large ship back to Europe. On his way back to Europe we met in Goa . . . Christopher [Borrus] was researching the system of Tycho and wrote in the field of mathematics, philosophy, and religious studies. He believed in the system of Tycho and he taught it in school. He also wrote a manuscript on Tycho's system, but his manuscript was lost when he was leaving China, due to a storm at sea. I wrote in Latin the words Christopher Borrus dictated to me, and then translated it to Persian. I hope I am forgiven for linguistic mistakes because my Persian language is very poor and I may not be able to convey his meaning. Therefore, I wrote in Italian as well, so you could find somebody in Persia who knows Latin languages, who would help you to consult the text.[8]

New Astronomy as a Missionary Tool

Della Valle gives only rough information about Borrus in the letter. But given that Borrus published his missionary experiences in transmitting astronomical knowledge to the Vietnamese, it is not too difficult to trace his career. Christopher Borrus, known also as Borri or Boro, was a missionary, a mathematician, and an astronomer; he was born in Milan in 1583 and died in Rome in 1632. He became a member of the Society of Jesus in 1616. In the same year, he was sent from Macao with Father Marquez, S.J., as one of the first missionaries to Cochin-China, where he stayed until 1622. On arriving in Rome, Borrus made an effort to bring to print accounts of his cross-cultural and intellectual experiences in that Indochina outpost, as well as the discoveries in natural philosophy that he was able to make during his travels.[9]

In 1628, Borrus published the work *Arte de navegar*, in which he presented various methods of navigation. Based on his knowledge of astronomy, he was able to make observations from different locations and thus contribute to cartography. He wrote articles on the use of the astrolabe, the measurement of longitude according to the clock, the constellations of stars as reference points, and the use of the celestial globe.[10] In 1947, one of della Valle's descendants, del Bufalo della Valle, donated to the Vatican Library a chart written and signed by Borrus.[11] It carried an early example of an isogonic chart, which later became a source for Athanasius Kircher and Edmond Halley.[12]

In 1631, a year before he died, Borrus published, in Lisbon, *Collecta astronomica*, a collection of essays that describes cutting edge topics in astronomy, from cosmological systems through explanations of comets. It also contains a

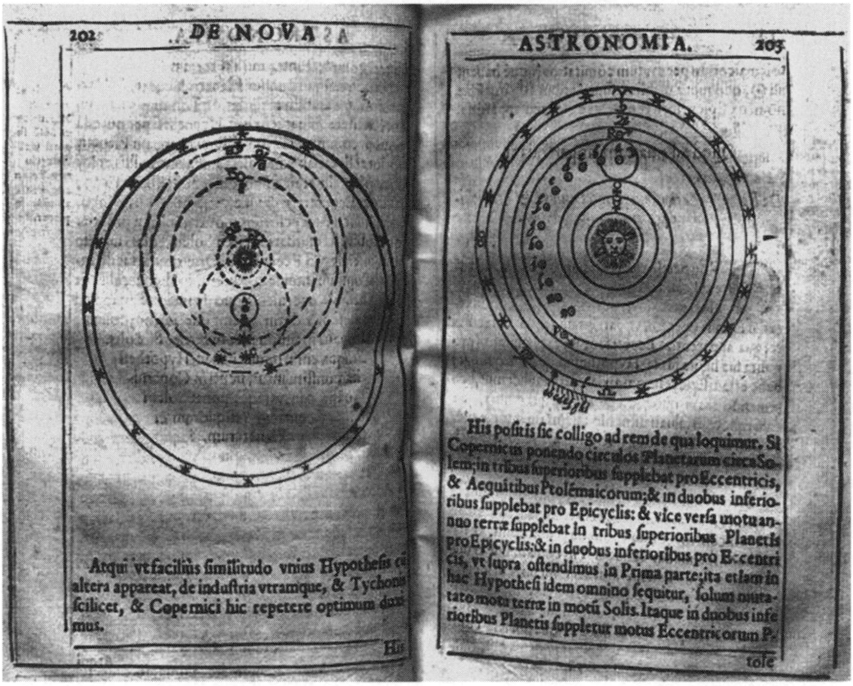

Figure 11. Two systems of the universe: Tycho (left) versus Copernicus (right). From Cristoforo Borri, *Collecta astronomica* (1631), IC6 B6472 631c. By permission of Houghton Library, Harvard University.

thorough presentation of the Tychonic system (figure 11). Here and there we find resemblances between the book he lost at sea and dictated to della Valle and the essay on the Tychonic astronomy.[13]

Although Borrus mentions the Copernican system as being one of the two ancient astronomies, along with that of Ptolemy, he mentions the Galilean discoveries as the most modern findings in the field, and he discusses Galileo's telescopic observations of the moon in addressing the question of "mountains" on the moon.[14]

> Someone will [perhaps] ask whether those protruding areas on the body of the moon, to which we give the name "mountains," are real mountains as on Earth. We will then answer together with Galileo Galilei: not only are they real mountains, but some of them are even higher than Earth mountains. To some, indeed, this will seem a difficult thing to say, and more difficult to prove; for who ever has flown up there to measure them? However, the proof is at hand

and quite obvious; for it is a particular ability of our [i.e., human] intelligence to get to places where the eye can never reach.[15]

We see that Borrus, although a devoted Tychonian, as expected for a Jesuit at that time, was still engaged with alternative astronomical theories and findings. His intellectual practice was not simply to introduce the Tychonic system, but also to introduce exotic cultures to his peers in Europe. In 1633, Borrus presented to the pope an account of Cochin-China and his experiences there. Apart from detailed descriptions of the climate, natural conditions, and politics, he noted two issues that are pertinent to our discussion.

First, Borrus saw the intellectual conditions in the kingdom of Cochin-China as made-to-order for missionary activities. The Confucian examination system, he noted, had a strong hold in the intellectual culture of Indochina. For examinations, the locals taught "the same sciences, using the same Books and Authors: namely Zinfa or Confus, as the Portugals called them being an author of as sublime and profound learning and authority with them as Aristotle amongst us, and indeed more ancient." He also acknowledged that "the ingenious invention of Printing was found out in China, and Cochin-China, long before Europe."[16] However, Borrus did not praise the Asians only for their arts. In a second treatise, dealing with the spiritual state of the kingdom, he related how he demonstrated to court astrologers and the king the superiority of European astronomy in predicting more accurately the solar and lunar eclipses and planetary conjunctions: "Now they know what the parallax is, which is the cause they are often deceived by, not finding the just time by their Books and Calculations." However, as in the case of the Jesuits in China, the aim of astronomical demonstrations was to promote conversions to Christianity. The Chinese literati were so deeply impressed that "it is not to be imagined how much Reputation this demonstration gained us among the Learned; insomuch that even the King's and Prince's Mathematicians came to us, earnestly begging we would receive them for our scholars; and upon this account the Fame of the Fathers was every-where so great, that not only our Knowledge in Astronomy, but our Religion was extolled above their own."[17]

As the letter to al-Lārī would indicate, Borrus's approach to proselytizing appealed to della Valle. He wanted to use the same method not only in converting the Persian astronomer he met in Lār, but also, as we shall see, in getting some specific objects in return.

Zayyn al-Dīn al-Lārī and the Perceived Persia

In his journal *De viaggi*, della Valle depicted Lār as an intellectual center. The local intellectual life charmed him. This little Persian town, he wrote, was "a leisure place, without court life, without ambition, without the distractions of urgent business, free from glamour and importunity of the soldiery, [and] was almost entirely given over to intellectual pursuit; and so successfully that nowhere that I have traveled in Asia, nowhere indeed in the whole world, have I found so many learned men, so many distinguished scholars as at Lār."[18] It was a local doctor who introduced della Valle to the circles of cultured men. He became accepted as a member of these circles and was entertained in people's homes; they helped him secure manuscripts to take back to Europe and gave him a great deal of valuable information on mathematical, astronomical, hermetic, and religious matters. A brilliant young mathematician and astronomer named Zayyn al-Dīn al-Lārī particularly impressed della Valle, who described al-Lārī as "an excellent astronomer and mathematician, supreme in the sciences not only in his place, but in all Persia and who is comparable to the best European astronomers . . . he [al-Lārī] wanted to learn Latin and to know all the astronomical charts and arithmetical figures."[19] Della Valle exchanged scientific objects with al-Lārī, such as manuscripts and books about mathematics, mechanical and hermetic curiosities, and versions of Scripture. The two men discussed astronomical questions. It seems that della Valle hoped that al-Lārī might accompany him to Europe, where "a man of his intelligence would surely exchange his Islam for the 'true faith.'"[20]

Unfortunately, we cannot learn anything about Zayyn al-Dīn al-Lārī's perception of the encounter, and we know of him only through della Valle.[21] Contemporary biographical collections do not list him,[22] and his name does not appear in catalogues of Islamic astronomical manuscripts.[23] Al-Lārī was an obscure figure and, unlike most other Muslims, was being subjected to Christian conversion attempts, and his meeting with della Valle raises some important points. While Jesuits traveled to the East to trade intellectual capital for souls,[24] merchants and travelers such as della Valle also traveled to acquire goods.[25] And the "good" that al-Lārī could supply was ancient manuscripts possibly containing remnants of pristine ancient knowledge.

Della Valle's travel in search of such knowledge was not exceptional. Other travelers in the Near East were on similarly motivated expeditions. Herbert Thomas, for instance, traveled in Persia and Goa in the 1620s, and his description of Lār may offer additional clues about the things that attracted della Valle and the context in which al-Lārī lived.[26]

At first, Thomas makes the comment that Lār is indeed a city that "pleads antiquity" and suggests that it is "the city which Ptolemy calls Corrha." He also tells us that Lār had a tradition of incorporating antique books and customs into Islam. He notes that "here are some proficient in Philosophy and Mathematics, the principle [*sic*] delight they take being in Astrology." In the nearby city of Shiraz, he found "a College wherein is read Philosophy, Astrology, Physics, Chemistry, and the Mathematics."[27] With Lār's rare combination of a long intellectual tradition and a symbolic status as a place that incorporated ancient Chaldean practices, Western visitors had high expectations of encountering manuscripts of pristine knowledge in this city.

Thomas refers to Babylon as the land of "Chaldea" and the location of Paradise, indicating a certain connection between "pristine knowledge" and the Chaldean sciences, and his travels aimed at recapturing these.[28] Thomas and other travelers, such as della Valle, emphasized the correlation between objects of "pristine knowledge" and Chaldea, on the one hand, and Near Eastern locales, on the other.

Collecting Ancient Manuscripts: Della Valle's Biblical Mysteries Tour

Thomas's idyllic Lār as the site of a mix of Chaldean Persian antiquity with Islamic intellectual culture is what, as we shall see, partly motivated della Valle's travels. The interest in Lār and in Zayyn al-Dīn al-Lārī far exceeded the urge to convert a Muslim to Christianity and was part of an overarching vocation.

Pietro della Valle (figure 12), born in 1586 to a distinguished Roman family, was educated in the classics and gained a considerable knowledge in Greek and Latin literature. He was admitted to the Accademia degli umoristi,[29] which was founded in 1603 as an institution for young, potential officers in the Church who sought protection under Cardinal Francesco Barberini, nephew of Pope Urban VIII. After completing his education, della Valle was eager to find out more about the Levant. In 1611, he took part in the expedition of the Spanish fleet to Barbary and was present at onslaughts against several pirate strongholds in the Gulf of Cabes, off the African coast. The port base was Naples, where he spent time and eventually made his decision to travel to the Near East.

Through the support of Schipano, della Valle embarked from Venice on June 8, 1614, destined for Constantinople. But he was not a pilgrim, as he claimed in the title of *De viaggi*. He did not go first to the Holy Land, but to Constantinople, where he spent a year studying Near Eastern languages and collecting

Figure 12. Portrait of Pietro della Valle (1586–1652) and the title page of his travel journal. From *De viaggi di Pietro della Valle il pellegrino* (Rome, 1650), 1416.14. By permission of Houghton Library, Harvard University.

manuscripts. The next point was Egypt and then the Holy Land, where he spent the least time; then he visited Syria, Mesopotamia, and Persia, ending at Goa. He traveled with a delegation of at least five members, including servants, painters, and translators, entailing substantial expenses; these expenses were covered by his patron, Mario Schipano, an enigmatic figure of whom we know little.[30] Schipano corresponded with della Valle during his travels and sponsored the travel in exchange for scientific objects—manuscripts, recipes for magic, curiosities such as mummies, local observations on celestial phenomena, and, finally, ur-text of the Holy Scriptures.[31] In a letter to Schipano, written in Constantinople and dated September 4, 1615, della Valle wrote: "I do not have some of the books yet, but I shall have all the others that you have told me about if they are to be found."[32] He also tells Schipano about the intellectual exchange with foreigners like himself and about the experience of purchasing manuscripts: "I will not refrain from saying to you that to obtain the Camus [Arabic dictionary], it has been necessary for me to play politics with the Turks, since an impertinent fool of a dervish threatened the seller, and insisted at all costs it should not be given to my men; he was suspicious that the buyer was a Christian, and said that it was wrong to give the books to Christians and their own works to *giaours*, as they call us contemptuously." Another manuscript "agent" whom della Valle employed in Constantinople was a Sephardic Jewish teacher, who taught him Arabic, Turkish, and Hebrew. The teacher once brought "many Arabic books which are for sale to show me . . . ; he also brought me a book on medicine."[33] However, della Valle's manuscript-buying agenda began to stray from language tools and turn more heavily toward natural knowledge. The letter to Schipano continues: "Not one of these [books offered] is what you ask for, but since [the topic] is medicine, as we have seen with the others, I have naturally arranged for the purchase of all of them to be negotiated."[34]

In Constantinople, della Valle was close to the French ambassador, Harlay de Cèsy, and the Franciscans—men with whom he observed cultural events and festivals and collected manuscripts. In September 1615, just before leaving Constantinople after more than a year's stay, he purchased antique manuscripts and medals.[35] De Cèsy escorted him to the port, secured his passage on one of the ships sailing for Egypt, and parted company, but not before pleading to be kept posted about the results of della Valle's search.

In Egypt, della Valle became an antiquarian, searching for ancient knowledge. Near the Pyramids of Giza, he conducted an excavation and found two portrait mummies, wrapped them in palm leaves, and carried them with him on his travels to take back to Italy.[36] He stressed the failure to purchase Samaritan Holy Scriptures in Cairo and seems to have been particularly interested in Moses and the language he used; he traveled to Mount Sinai where "the Holy Scriptures was first written."[37]

Della Valle's interest in ancient manuscripts, Chaldeans, and Moses and Mount Sinai had particular motivations. While at first he looked for Arabic dictionaries and medical books, during the rest of his travels, and especially after 1616—either attracted to biblical sites or responding to direct instructions from Schipano—the search focused on other objects. Ancient versions of the Scriptures, especially those regarding the small religious community of Samaritans, fascinated him and determined his itinerary. While searching in Nablus and, later, near Damascus, he looked for Samaritans' ancient Holy Scriptures. On this topic, he wrote to Schipano from Aleppo on May 1, 1616.

> As well as the delight I took in seeing their gardens and houses . . . I was in heaven over their richly illuminated Samaritan manuscripts, and also their synagogue. I was also made very happy to see in the house of one of their *haham*, or wise men, four books of the Sefer ha-Thorah [the Scroll of the Torah] of

> the Samaritan manuscripts I had been hunting for so diligently. These were all very ancient books, all written in Samaritan on large sheets of vellum . . . To conclude, I was so successful with a little money and through the diligence of my Jewish interpreter, I took two of these books and I wished to give [one] to my friend de Cèsy, the French Ambassador in Constantinople, who wanted this version and to whom I have already sent it.[38]

Della Valle's excitement suggests both a need and a quest, motives that he shared with Schipano and de Cèsy: he wanted various ancient versions of the Scriptures that preceded the Hebrew Bible, as well as, of course, the Vulgate and Septuagint. Furthermore, he launched a critique of the Vatican institutions' policy that restricted access to collections of books in the Vatican Library and forced scholars to turn elsewhere.

> I am sure that no [ancient Scriptures manuscript] like these can be found in Italy, not even in the Vatican library. Some have advised me to give it to the Vatican library, as something rare; but rather because it is rare, I have resolved (and think perhaps it better) to keep it by me while I live; not least because, in the Vatican library, to which few people have access, among such a multitude of books it would probably be buried and scarcely known, whereas in my hands it will be continuously exhibited, to the public benefit of every person of talent wishing to make use of it, and to study it. I intend this to be so with all the curious things I have found and acquitted through my labors.[39]

But della Valle was not traveling and collecting items simply for a prospective cabinet of curiosities, which he intended to build on his return to Rome. His interests extended from objects of natural philosophy and alchemy to ancient biblical texts and ancient personas and their tombs. One figure that occupied his mind, apart from Moses, was Job. In visiting Hums (Homs), in northern Syria, della Valle looked for Job's tomb. On May 30, 1616, he wrote: "I was told that the local Christians venerate, I do not know whether in Homs or Hamath, a certain memorial of Job, and they hold to the opinion that he lived in those parts. I did not see it, as I had not been told in time. I also suspended judgment as to the truth of this because I suspected the town was a little too far north to have been Job's own native place."[40]

The Greek philosopher Pythagoras was another figure of particular interest to della Valle and a further motivation for his travels (figure 13). In India, for instance, an old Brahman showed him a translation of the works of Pythagoras, whom he alleged to be one and the same as Brahma.[41] Della Valle was interested

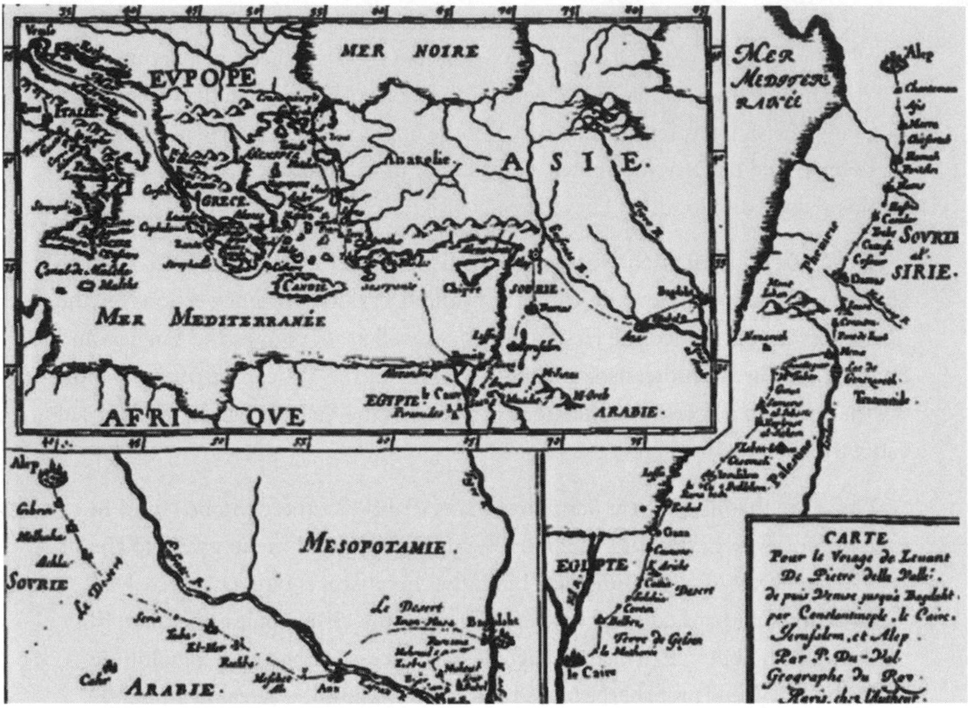

Figure 13. The Pythagorean itinerary of Pietro della Valle, as illustrated in his journal. From *De viaggi di Pietro della Valle il pellegrino* (Rome, 1650), I:4. By permission of Houghton Library, Harvard University.

in Pythagoras not only as a Greek philosopher, but as a role-model philosopher who traveled in the Near East and collected pieces of ancient wisdom from Hebrews, Chaldeans, and Indians. Moreover, sympathizers with Copernican cosmology in the early seventeenth century were referred to as "Pythagoreans," and Pythagoras and his sources came to be identified as a historical fount.

The Non-Tychonic Assertions in della Valle's Letter to al-Lārī

Della Valle, then, was a polymath, devoted to collecting manuscripts of natural philosophy and the Bible, exploring biblical tombs, and making casual observations in astronomy. His preexisting interests in these fields may have set the context for the letter to al-Lārī. With this in mind, we return to della Valle's introduction in the letter. In nonfluent Persian, della Valle tells al-Lārī about the new European discoveries in the field of astronomy.

> In Europe some people have been writing about the "long eye" [*cheshmak de-rāz*], the object that can see far away [i.e., the telescope]. People have written [about] how to produce it and all of its secrets and benefits. And they have written about the beaming objects [satellites of Jupiter] in the sky, which are very far away from us. Especially the two big ones that were seen five to six years ago. There have also been various other books in the field of astronomy. And when I go back to my own country, with the will of God I will send them all to Mullana [i.e., to al-Lārī], including the book of Tycho Brahe and also the astronomer of the Emperor Rudolf II, Kepler, who is the most famous astronomer these days. At the end of his life he wrote a book that is very valuable and that made him very famous. In his work he presented the structure of the universe in a different way than people before him.[42]

Some details are noteworthy here. Della Valle mentions the cutting edge of the field of astronomy, including mention of a Copernican such as Kepler. Moreover, while blurring the profound disagreements between the Copernicans and Tycho and the Jesuits, he brings up the findings of Galileo and the telescope. Yet he mentions neither Copernicus nor Galileo by name. Later in the introduction, della Valle implies that their works are inconsistent with the work he translated from Borrus: "However, nowadays I have heard about the many up-to-date and better things in astronomy in Europe since I wrote this piece. And these new works are against Tycho's system. I want you to know some of the sentences that make sense. I wish to serve you, and [that] you would at least know this system. In order to facilitate your understanding, I also put a diagram of Tycho's system, which shows that earth is the center of all the stars" (figure 14).[43]

Della Valle's letter, then, avoids mentioning Copernicus and Galileo explicitly. In *De viaggi* we find no trace of the names Copernicus, Galileo, and Kepler, and no indication that della Valle transmitted information about Galilean and Keplerian discoveries to any locals. However, he made an implicit contrast between the Copernican astronomy and his rendition of Borrus's words. Recall that Borrus's *Collecta astronomica* engaged with cosmological controversy, and apparently, as a hidden agenda, della Valle as a translator took the liberty of "self-censoring" Borrus and omitting the names of Copernicus and Galileo. Thus, not only was della Valle familiar with the new and quickly rising astronomy, but, more important for our further reading of the letter, he was aware of the persecutions that Copernicans experienced in Europe, and especially the 1616 trial of Galileo and the inclusion of Copernicus's *De revolutionibus* in the index of prohibited books.

Della Valle was writing the letter hundreds of miles away from Europe, but the

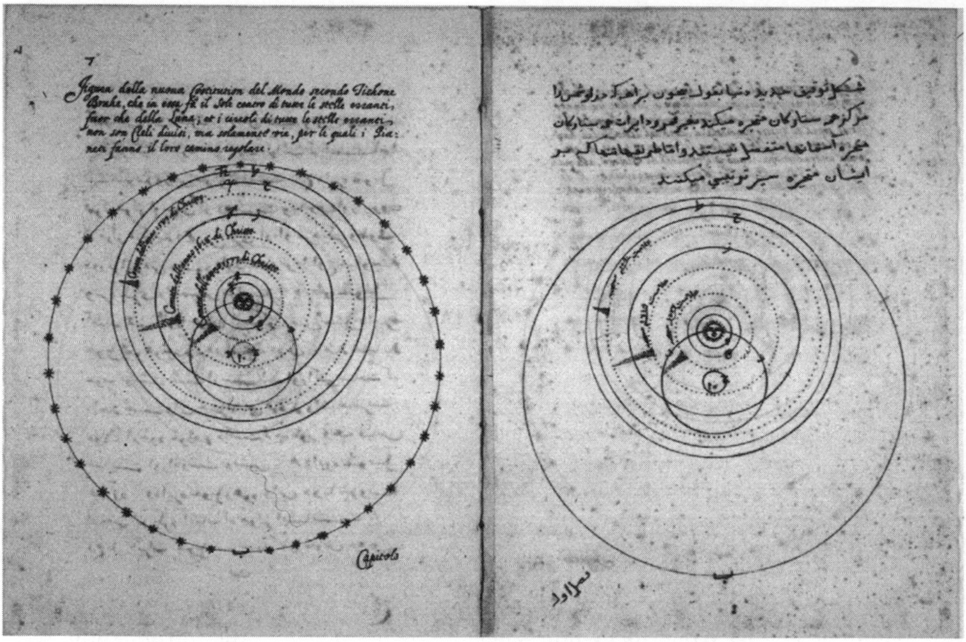

Figure 14. Della Valle's diagram of the Tychonic universe, in Italian and Persian. From Pietro della Valle, "Letter to Zayyn al-Dīn al-Lārī," Persian Collection, MS 9, Vatican Library.

fear of the Inquisition echoed in the Portuguese colony. The Inquisition in Goa had a local and particular agenda in persecuting local converts who kept their Hindu practices and beliefs or to protect them from Protestant propaganda; however, it still remained an extension of the European Inquisitions. In 1606, the fifth Concilio Provincial passed the following resolution:

> This Sacred Council directs, under pain of excommunication *latae sententiae*, all the captains of the fleet, soldiers and Christians of any rank or condition residing in this state, if they come into possession of any book brought from Dutch or English ships, or those of any other foreign nation, navigating in these parts, and in whatever language the book may have been written, not to read or give it to others . . . even though the titles may appear pious and devout, for the very reason that, as is the way of heretics, under such titles a lot of false and pernicious doctrines are contained against the truth and the purity of the Holy Catholic Faith.[44]

In Goa, apparently, della Valle had every reason to be cautious and to conceal his sympathy toward the Galilean findings that shook the Tychonic system. However, it is not only his knowledge of and sympathy toward Galileo that he had to conceal, but also his social associations and intellectual vocation.

Della Valle's Hebraist Take

Even though della Valle's letter to al-Lārī was a translation of Borrus's work on Tycho, the introduction, as described above, mentioned other astronomical findings and implied that they were incompatible with the Tychonic system. Moreover, at the end of the letter, the author breaks off from the translation and adds a crucial clue: "This is the abstract of the book of Christopher Borrus, which I translate. It has made me content, and also I agree with it. But, certain verses of Job the prophet raise a little doubt. The Book of Job was translated to Latin and was in the hands of observant believers, but the real Book of Job the prophet is in the language of Hebrew and Chaldean."[45]

As an aside, we see that della Valle shows familiarity with contemporary theological matters surrounding the debates over the Copernican system. One of the first works to reconcile the Holy Scriptures and Pythagorean cosmology was by the Hebraist Diego de Zuñiga, *In Job commentaria*, an exegesis on the Book of Job. In the commentary on Job 9:6, which discusses the omnipotence of God, de Zuñiga takes the liberty of engaging with contemporary astronomical theory. The verse mentions that God "shaketh the earth out of her place, and the pillars thereof tremble" (*qui commovet terram de loco suo, et columnae eius concutiuntur*),[46] and for de Zuñiga, it was a scriptural clue to the truthfulness of the Copernican theory. De Zuñiga writes: "for those to whom this particular passage seems difficult, it might be illustrated by the Pythagorean doctrine imagining the earth to move by its own nature; by no other means can we explain the motion of the planets so greatly differing in speed and slowness . . . In our time, Copernicus announced a motion of the planets in accordance with this ancient opinion."[47]

Della Valle's comment may have been echoing sympathizers of the Copernican cosmology who used de Zuñiga's work. The statement about the ur-text of the Book of Job turns the Scriptures into a judge in the Copernican-Tychonic controversy. It acts as a crucial piece of the puzzle that shifts our investigation from Goa back to Europe.

A Larger Picture: Hebraists and Hermeneutic Struggles

Robert Westman suggests that de Zuñiga's commentary belonged to an intellectual tradition evident in the liberal Hebraist circles of Salamanca. The commentary resulted from contemporary Catholic trends to reform the Vulgate by comparing it with the more ancient Greek, Hebrew, and Chaldean versions.[48] Moreover, in addition to the return to ancient versions of the Bible, new commentaries such as de Zuñiga's, which addressed the matter of a changing cosmology, also incorporated a hermeneutic device: the "principle of accommodation."[49] A wider interpretation suggests that the Scriptures were written in a mundane human language so as to simplify complex concepts and gradually transform pagans into monotheists. However, the two trends in scriptural commentary seem askew. On the one hand, those who used the "principle of accommodation" believed that the "secrets of nature" were embedded, albeit obscurely, in the Scriptures. For them, scriptural text was fixed, so they had to decode its implicit messages to make it less obscure. On the other hand, those who looked for biblical texts other than, and prior to, the Vulgate believed that the translation processes had corrupted some of the secrets of nature and that the "original" texts of the Scriptures explicitly mentioned these secrets.

Thus, contemporary thinkers used both types of hermeneutics as strategies and tools in their struggle to reconcile theology and Scripture with the new findings and theories in cosmology. However, the trend of consulting ancient versions of the Bible faced an obstacle when the Council of Trent resolved that the Vulgate was the only recognized version for the Catholic Church.

Della Valle's arguments regarding the Scriptures have yielded traces of a contemporary hermeneutic debate. To give a sense of place to that debate, we turn to Italy, which was in a favorable position to become a center of the new movement in Hebraist studies. The elements particular to Hebraist learning were present: an Arabic-Spanish Jewry, a developing and spreading craft of Hebrew printing, and an interest among Christian scholars in the Cabala. Christian scholars of Hebrew in the sixteenth century quickly directed their linguistic skills toward biblical critique.

In Genoa, in 1516, the Christian Hebraist Agostino Giustiniani (1470–1536), Bishop of Nebbia, published the *Psalterium polyglottum*, containing the Hebrew, Greek, Chaldean, Arabic, and Latin texts of Psalms, together with Latin commentaries.[50] Santes Pagninus, another Italian Hebraist, made a new Latin translation of both the Old and the New Testaments that found rapid acclaim among scholars. The Council of Trent tried to stop this trend, declaring that the only acceptable biblical text was the Vulgate. However, it could not stop the printing of the *London Polyglot* (1647), which used Santes Pagninus's new translation as the Latin text.[51] Publications of other commentaries, however, challenged the attempts to continue the liberal Hebraist trend—for example, the Jesuit François Vavasseur's *Iobus: Carmen Heroicum* aimed to correct the damage done by de Zuñiga, by relying solely on the Vulgate.[52]

The Radical-Liberal Hebraist Take: The Search for the Lost Book of Job

Victor Navarro Brotóns supplies ample evidence for the way in which the liberal Hebraist de Zuñiga defended his commentary on the Book of Job from allegations that it was heretical—because he had consulted both other versions of the Bible and natural philosophy. "With arduous labor and great diligence," de Zuñiga writes, "I have devoted myself to the study of letters, having learned Latin, Greek, Hebrew, Chaldean and Italian well enough. I have tackled all the arts and sciences . . . I have read all of the sacred books at least twelve times, in their original languages: Hebrew, Chaldean, and Greek, using the best guides, with the result that the sacred books are so familiar to me that that there is no passage, whether written in Hebrew, Chaldean or Greek, that I cannot explain at sight, in such a way that my explanation cannot be criticized by any learned man."[53]

Apparently, de Zuñiga, who was the first to make the connection between the Book of Job and Copernican cosmology, was the source for della Valle's remark about Job in his letter to al-Lārī.[54] Della Valle not only repeats arguments made by de Zuñiga, that he needs to consult the Book of Job in other languages, but actually takes the arguments one step further. Della Valle writes: "One should look for the original Book of Job in the original language. Therefore one should look for the saying of Job in the original language and what power of benefit his saying has. So if one would look at the original piece that is the statement of Job himself, that is good! But if it is the statement of God to Job then it is a command of God and we cannot say anything against it."[55] Della Valle adds here a new hermeneutic argument regarding textual voice. Commentaries on the Book of Job traditionally divided the work into three different speakers: God, Job, and Job's fellows.[56] For della Valle, if the verse was a statement of Job, it would still not entirely shake the Tychonic cosmology. But if it was God's command, there would be no choice but to accept the possibility of the earth in motion. Moreover, della Valle's letter repeatedly asserts the need to find the Book of Job in its "original" language.

> . . . we do not have the Book of Job in Hebrew and Chaldean that could point out the cosmological truth. With God's will these original texts would someday resurface from the treasury in the basement of the Vatican . . . for the time being we could avoid relying on a sole source like the Vulgate [by] consulting commentaries on the Book of Job in Hebrew and Chaldean.[57]

Some items here are worthy of attention. Della Valle emphasizes that lacking the original Book of Job, we should rely on the commentaries of Hebraists such as de Zuñiga, who can fill the gap through their skills in Chaldean and Hebrew. More striking, della Valle asserts that the original Book of Job in Chaldean and Hebrew could actually be found in the Near East.[58] Thus, the type of treatment della Valle meant to give to the Book of Job accorded neither with the "principle of accommodation" nor with Hebraist tradition. It was a new concept: that the explicit secrets of nature and Pythagoreanism could be found in lost scriptures, outside Rome, in the ancient Near East.

Della Valle went a step further: he disregarded the Latin, Greek, and rabbinical Hebrew texts of the Bible. Thus, the Book of Job, for its language and myths, was considered an early book of the Bible, and della Valle's search for the "original" text of the Book of Job was actually a search for pre–Second Temple scriptures, which could be found among Samaritan communities, in the Chaldean language, and in the Chaldean Nestorian church.

It was not della Valle who invented the argument about lost originals. Avraham Ibn-'Ezra (1092–1167), an astronomer and exegesis writer, stated in his commentary on the Book of Job that, with regard to its difficult language, "it seems probable to me that it is a translated book and that this is why, like all translated books, it is difficult to interpret."[59] Ibn-'Ezra's commentaries were readily available to early sixteenth-century Christian exegetes, in both Hebrew and Latin.[60]

De Zuñiga, and consequently della Valle, saw the widely available versions of the Book of Job as imperfect translations of the original book written in Chaldean in a very early phase of the writing of the Bible. Moreover, according to later tradition, the Book of Job captured Mesopotamian myths that prevailed in the Chaldean kingdom in the eighth century BC, a time that paralleled the period of the First Temple.[61] Della Valle wanted to find the Book of Job in the Chaldean language, as he requested of al-Lārī: "If you find an educated Jew in Lār ask him about the *word* in the Book of Job the prophet. However, the Jews nowadays go in the wrong religious way. But, [I ask you to ask the Jews] because the books of antiquity, meaning the books of the saying of God, which were written before Jesus, among them the Book of Job, all [are] written in Hebrew and Chaldean. And the Jews who have the book of the religion are correct in believing their books."[62]

Della Valle mostly traveled in the areas of Mesopotamia, where the ancient Chaldean-Babylonian kingdom was located, so why did he go south all the way to Goa? The west coast of India was home to several flourishing Chaldean-Christian communities, which the clergy of Goa tried to incorporate through Latinization; the Jesuits followed a similar, if more moderate, policy.[63] Thus, della Valle's search for the Chaldean-Hebrew original of the Book of Job most probably did not begin or end with requesting al-Lārī to ask among the Jews of Lār. His fascination with the Samaritans, who were remnants of the Israelites' exile to Chaldean Babylon (though they did not preach the Book of Job), and with the Nestorians (one of whom he married) and the Mesopotamian and Persian Jews, as well as his continuing travel along the west coast of India—all indicate that a primary motive was the search for ancient books in general and the Chaldean-Hebrew "original" Book of Job in particular. Della Valle treats the Book of Job as a source through which he could find out the "true" cosmological system.

Once again we face those divergent texts—*De viaggi* and the letter to al-Lārī. We have no direct evidence in *De viaggi* of a motive for a Chaldean search, because the issue at stake was highly sensitive and under the scrutiny of the Inquisition. As mentioned above, della Valle took the liberty of discussing in the letter issues of textual origins, though cautiously and in a non-European language. The letter's format, language, and destination mark the political and literary boundaries of the European intellectual market.

The Neapolitan Reconciliation: A Source of della Valle's Quest

Della Valle's search for the ur-text of the Book of Job was not a solitary endeavor. His sponsor, the Neapolitan physician and natural philosopher Mario Schipano, provided him with an intellectual frame of reference anchored in Neapolitan intellectual culture. Schipano's intellectual pursuits may thus become clear in the Neapolitan reactions to the Galilean affair of 1616.[64] In the inquisitorial decree of 1616, we find the connection to Naples.

> This Holy Congregation has also learned about the spreading and acceptance by many of the false Pythagorean doctrine, altogether contrary to the Holy Scripture, that the earth moves and the sun is motionless, which is also taught by Nicholaus Copernicus's *On the Revolutions of the Heavenly Spheres* and by Diego de Zuñiga's *On Job*. This may be seen from a certain letter published by a certain

> Carmelite Father, whose title is *Letter of the Reverend Father Paolo Foscarini, on the Pythagorean and Copernican Opinion of the Earth's Motion and Sun's Rest and on the New Pythagorean World System* (Naples: Lazzaro Scoriggio, 1615), in which the said Father tries to show that the above-mentioned doctrine of the sun's rest at the center of the world and the earth's motion is consonant with the truth and does not contradict Holy Scripture.[65]

The treatise mentioned in the decree, in which Foscarini was eager to reconcile the Holy Scriptures with the Copernican system, was written and printed in Naples—and was the last straw that led to the inclusion of Copernicus in the index of prohibited books. However, Foscarini's textual analysis of Scripture argued differently from de Zuñiga's. While the Spanish Hebraist looked for different versions of the verses of the Book of Job, Foscarini extensively used the "principle of accommodation" and exclusively relied on the Vulgate. Foscarini complained that "we are not to have so high a respect for the Antiens [Ancients], that whatever they assert should be taken upon trust, and that Faith should be given to their saying, as if they were Oracles and Truths sent down from Heavens."[66] Although some of Foscarini's arguments echoed at Goa, della Valle and his sponsors took a different direction from Foscarini and the "principle of accommodation."

Foscarini's treatise was written in Naples in 1614 and published in 1615, when Schipano sent della Valle to the East. Thus, for Schipano, as for della Valle, Foscarini was apparently a source of the idea that Scripture and the Copernican system could be reconciled. However, as the Church rejected Foscarini's treatise and issued the inquisitorial decree of 1616, Galileans who still wished to reconcile the Pythagorean system with Scripture looked to other strategies, such as the search for the "original" Book of Job in the Chaldean language.

The connection to Naples requires explanation. In fact, debates in Naples were not a matter of wide interest, but took place among exclusive circles of highly motivated scholars. Beyond mere philosophical connections, della Valle was affiliated with a cluster of Neapolitan scholars, followers of Pythagoras, scriptural commentators, and travelers to the Levant.

Neapolitan Hermeneutics and Secrets: The Pythagorean-Hebraist Agenda

The correlation of Hebraism and natural philosophy went beyond biblical matters of legitimacy.[67] It represented a well-rounded perception that Pythagorean and hermetic philosophies synthesized ancient Hebrew, Chaldean, Egyptian, Persian, and Hindu knowledge of nature. Renaissance men saw Pythagoras as living in a period of rather easy cultural exchange: he traveled to Egypt to study the virtues of numbers and geometry, and then to Babylon, where the Chaldeans taught him the course of the planets. Pythagoras then wandered in Persia and India, and then back to Calabria. Early modern scholars perceived his itinerary as potentially recovering a synthesis of ancient wisdom.

Other Hebraists and cabalists followed the same line of thinking and identified ancient Jewish wisdom as the source for Pythagoras. Johannes Reuchlin, a prominent sixteenth-century Hebraist, stressed that all great ancient wisdoms came from a Jewish spring. He argued, for instance, that Pythagoreanism was entirely based on Jewish number codes, which were in turn predicated on a most exciting use of Hebrew.[68]

Such a connection between ancient Near Eastern wisdom and Pythagorean heliocentric cosmology was made in Naples, the intellectual anchor for della Valle's travels. The letter to al-Lārī shows an engagement with Neapolitan interests, particularly with a cluster of Galileo sympathizers. Given that Foscarini's work was the most renowned defense of Galileo, one might think it a suitable source and an instigation for della Valle's travel. However, on the basis of the hermeneutic nuances and the lack of evidence for any mutual personal connections, Foscarini does not seem to be the source.[69] Nevertheless, Foscarini's central role in the "Galilean affair" of 1616 acted as "a foam above the stormy intellectual water" of 1610s Naples. We now dive into that water.

The "Galilean affair'" started with Galileo's *Letters on Sunspots* (1613), in which he brought up the matter of the Copernican system and for the first time endorsed it unequivocally in print, predicting that it would soon be universally adopted. Soon thereafter, opposing camps of theologians, mathematicians, and philosophers within the Church struggled over the question of Copernican cosmology. In 1615, Galileo wrote his *Letter to the Grand Duchess Christina*, in which he argued that the Holy Scriptures should not be a source for natural philosophy. In the years 1613–16, discussions over the compatibility of the Copernican system with the Holy Scriptures arose but were not yet necessarily connected to Galileo's letter, which was not widely circulated. The locus of the proponents of the compatibility of the Scriptures with Pythagorean cosmology was Naples, where theologians and natural philosophers engendered a strong feeling for the Galilean camp. By the beginning of February 1616, Galileo seems to have felt the need to visit the Neapolitan center and, most probably, to see Foscarini and others; he wrote a letter to secure permission from the Grand Duke for a visit to Naples. However, the intra-Church debate over the Copernican system had ended with the consulars of the Congregation of the Index deciding to condemn

Pythagorean cosmologies and three Copernican works in the index. Galileo then became aware of the threat that a spontaneous visit to Naples would pose to the Church. He had to withdraw the plan to visit, "because of the bad weather and roads."[70] And indeed, Naples turned out to be a dangerous locale for Galileo sympathizers.

Foscarini died before he could be prosecuted, and his publisher had to run for his life. From here onward, others hid their engagement with Pythagoreanism. However, Foscarini was but one theological defender of Galileo who stressed a reconciliation of Pythagorean cosmology with the Scriptures through the "principle of accommodation." Others in Naples worked on the relationship between the ancient Scriptures and Pythagoreanism—not just the connection to Jewish wisdom, but the historical belief that Pythagoras had contributed to the chain of transmission of the revelation. One of these thinkers was Tommaso Campanella, from Calabria, the home of Pythagoras. He was educated in Naples, where he was associated with Giambattista della Porta and other advocates of magic. For political conspiracy against the Spanish rule, he was charged and imprisoned by the Inquisition. In prison he wrote a defense of Galileo, *Apologiae pro Galileo*, written as early as 1616 but not printed until 1622 (figure 15), in which he cited sources claiming that Pythagoras was Jewish.[71] Campanella argued that when God revealed the moral laws to Moses, he also revealed the secrets of nature, which were transmitted in extra-scriptural traditions such as the Cabala. Campanella gives some detail concerning the chain of transmission: "Galileo's theory of the motion of the earth, of a central Sun, and of the systems of stars with waters and earthly elements is indeed an ancient conception. It comes from the mouth of Moses himself, and then Pythagoras," who promulgated it to the Gentiles.[72] Relying on Pico della Mirandola, Campanella stressed that Aristotle, and as a result Thomism, deliberately broke from the ancient revelation and distorted human understanding of the universe. Thus, in presenting Aristotelianism as distanced from the true revelation and in recapturing the sources of Galileo and Pythagoras, Campanella suggested that the theory of heliocentrism corresponds to the revelation of Moses on Mount Sinai.

Campanella's hermeneutic approach was different from Foscarini's and de Zuñiga's. In using the principle of accommodation, Foscarini could manipulate the interpretation of the fixed text in the Vulgate. De Zuñiga used the same principle of accommodation, but preferred to consult other sources, not just the fixed text of the Vulgate. Campanella, however, presented something completely new: all versions of the Bible are only corrupted derivatives of lost ur-text. Thus, given that the Church argued for the incompatibility of the current versions of the

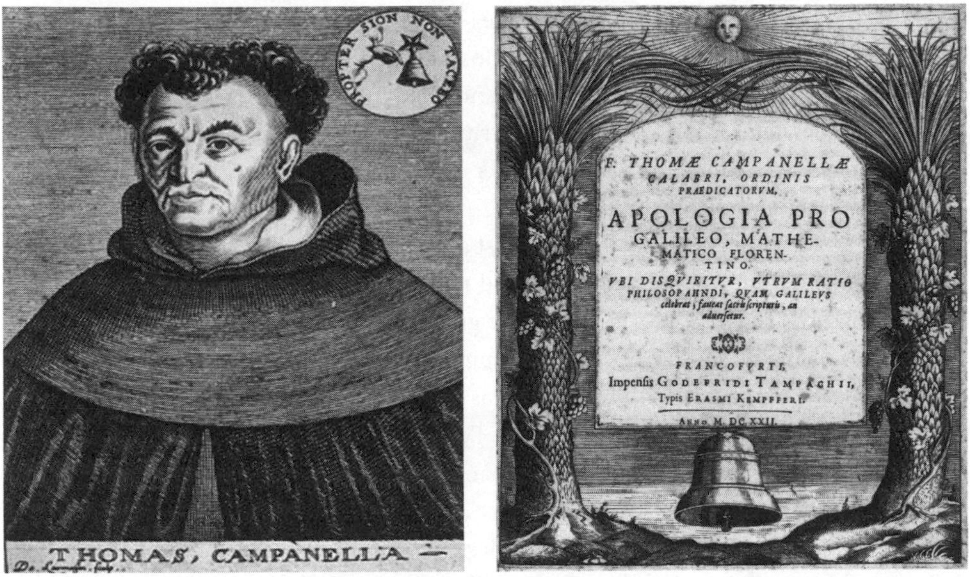

Figure 15. Portrait of Tommaso Campanella (1568–1639) by Nicolas de Larmessin, ca. 1650–60, and frontispiece of *Apologia pro Galileo*, published in Frankfurt in 1622. Nicolas de Larmessin II was a member of a family of printmakers and booksellers. The family had its own publishing house in Paris, designing and printing books, prints, calendars, and other popular works on paper. Campanella moved to Paris in 1634 and stimulated great intellectual and cultural interest, which still vibrated a decade after his death when de Larmessin made his portrait. Left, from *Les avgvstes representations de tovs les roys de France depvis Pharamond ivsqv'a Lovys XIIII . . .* (Paris: Bertrand, 1679). Right, from Campanella, *Apologia pro Galileo*, IC6.C1513.620d; by permission of Houghton Library, Harvard University.

Scriptures with Pythagoreanism, Campanella creatively argued that the revelation of Moses was incompatible with the current versions of the Scriptures. And therefore the search for pristine scriptural knowledge might lead to a reconstruction of the "revealed" secrets of nature, a process that Copernicus and Galileo had started by exploring the "book of nature."

For Campanella, it was something of a logical next step to locate pristine scriptural knowledge—the secrets of nature—so as to corroborate Copernicus and Galileo. He marshaled evidence to argue, at the end of *Apologia*, that the ancient Pythagoreans of his native Calabria derived their heliocentrism from Jewish sources. Moreover, Campanella stressed that even if Pythagoras was not Jewish, one could still retrieve evidence of the "true history" of the connection between the revelation of Moses and Pythagorean cosmology through the writings of

"Egyptian priests and of the Jews in Judea bordering upon Syria and Egypt. From them we hear both the Law and the philosophy of waters, mountains, and earths in heaven, of mountains in the Moon and similar things."[73] Consequently, della Valle's search for Samaritan and other writings in Egypt, Syria, and the Holy Land resonated with Campanella's arguments in favor of heliocentrism, Copernican cosmology, and eventually Galileo.

Campanella's interest in the Near East and the Ottomans went beyond merely locating pristine knowledge. It included the politics of apocalypse. Campanella was imprisoned in Naples in 1599 on the grounds of conspiracy against the Spanish government. The details of the conspiracy are striking. In June 1599, an Ottoman fleet commanded by Mūrat Reis was anchored near Reggio Calabria. Mūrat Reis was actually an Italian, originally Scipione Cicala, captured as a boy and recruited to the *devshirme*, the Ottoman institute for training captive boys as bureaucrats and soldiers. The now successful Mūrat Reis had berthed at Reggio Calabria merely to see his mother in Messina. But Campanella and his partners urged Mūrat Reis to invade Calabria so as to intensify the apocalyptic process, which, they hoped, would cause the pope to flee to a utopian retreat called *civitas solis*, the "City of the Sun."[74]

Thus, the intellectual inclinations of Tommaso Campanella, a Neapolitan esoteric theologian, and his interests in the Near East, resonated conceptually with della Valle's quest. However, for della Valle, Campanella was not only an intellectual source but also the engine for his travel.

Building a Galileo-Campanella-Schipano Connection

Scholars have discussed the fine threads and personal links that connected the Galilean affair to the Near East. Mario Biagioli mentions that Galileo's patron, Sagredo, was sent to the Levant in 1608 as the Venetian ambassador. He served for three years in Aleppo, during which time he corresponded with Galileo, asking him to send articles to be used in gift exchanges. Sagredo was conducting astronomical observations and was in contact with local Muslim scholars; he also corresponded with Jesuits in the Far East.[75] Thus, the arrival of della Valle at Aleppo six years after Sagredo's diplomatic mission had ended was not in itself an unusual itinerary for a Galilean. However, della Valle's agenda was different.

The specific motives mirrored the personal connections between Galileo, Schipano, and Campanella. Schipano has an elusive identity. He is virtually nonexistent in the documentation concerning the Galilean affair. Accounts of the history of medicine and scholarship in Naples provide no information on him.[76]

But despite the dearth of evidence surrounding Schipano, we do have a sample of his handwriting in a manuscript in the Lincei Archive. The guest book of the Lincei Society lists the society's candidates who visited its center in Rome. Each candidate wrote a short sentence, in the way of a *vita*. Schipano's item reads: "di Napoli, medico philosuphus non arabicae linguae ignarus."[77] Schipano visited there sometime in 1618, after being nominated to the society by Fabio Colona. However, for a mundane reason, Schipano was not admitted to the society: Federico Cesi, who was in charge of the admission process, was sick at the time and could not interview the prospective fellow.

Schipano worked as an informal member of the Lincei circles in Naples. When Federico Cesi and Fabio Colonna became interested in opening a branch of the Lincei Society in Naples, they consulted local intellectuals, one of whom was Schipano.[78] The Lincei, by and large, expressed support of Galileo and especially backed his right to intellectual freedom. Even on the sidelines of the Lincei, Schipano was familiar with Galileo and closely followed the political storm. Galileo and Schipano were indirectly connected. Fabio Colonna, a leading botanist, was an active member of the Lincei Society. He tightened relations with Schipano, with whom he mainly discussed botany and judicial astrology in their uses for medicine. Moreover, as a leading member of the society, he eventually established a long-distance relationship with Galileo as part of a formal academic correspondence on various disciplines and on matters concerning Galileo's research methods.[79]

Colonna was a conspicuous student of astronomy among the members of the Neapolitan Lincei Society. He cultivated the study of Galilean discoveries to such an extent that they became renowned in Naples.[80] His interests went in other directions as well—natural philosophy, alchemy, and astrology. He was excited about the telescope, which stimulated him to produce a microscope for his experimental research in botany.[81] Colonna, ostensibly, was an intermediary between Galileo and Schipano.

We also learn of the secretive nature of Neapolitan Pythagorean societies. Schipano was well-grounded in various realms of natural philosophy, but with a Pythagorean twist. He was mentioned as a magus, astronomer, astrologer, physician, pharmacist, botanist, and natural historian. He led a Pythagorean circle of natural history and botany, which looked up to Galileo's scientific persona. Having witnessed the persecution of its idol, the circle became a relatively secret order: followers refrained from conducting public gatherings and making statements, preferring intimate intellectual exchanges. Moreover, the botanical Lincei circle of Schipano was secretive not only in conduct but also in the sense of having great interest in the secrets of nature.[82] Various travelers were sent as messengers on behalf of secretive Neapolitan academies to discover natural secrets or books on the secrets of nature.[83]

Widening our perspective on Naples a bit further, we find that the secrecy of Schipano and della Valle was not just a result of debates over the secrets of nature or even the caution required to avoid persecution by the Spanish. It was also a reaction to attempts to institutionalize Neapolitan intellectual culture and bring it under the firm structure and dogma of a newly reformed university. Pietro Giannone's contemporary account of Naples tells us that in the early seventeenth century, the government of Count Don Pedro Fernandez de Castro invested in new buildings, and in 1616 the university was reopened with a new curriculum and regulations modeled on those of the University of Salamanca. The count provided the university with a copious library and prescribed the methods of collecting, cataloguing, and preserving the books. However, Giannone also mentions that some central figures refrained from formal university participation and, "leaving the beaten path, went the right way to work, and at length gave light to posterity to follow their footsteps . . . [among them] Mario Schipani, an able physician, and an intimate friend of the great virtuoso and traveler, Pietro della Valle."[84] Schipano and some others, as Giannone notes, chose to create an alternative circle with its own book collection, to which della Valle's travels and purchases contributed.

A more personal connection tightened the relationship between Schipano and Tommaso Campanella. As noted earlier, Campanella wrote the apologia on Galileo during the years 1614–16 as a prisoner of the Inquisition in Naples. He consulted a great number of books, which were brought to his cell by Neapolitan friends,[85] among whom were Fabio Colonna and Mario Schipano.

Luigi Amabile's voluminous accounts of Campanella mention that Schipano's connection to Campanella went back to Tommaso Campanella's uncle, Giulio Campanella, who, like Schipano, was a teacher outside the university institution.[86] Elsewhere, Amabile mentions the friendship between Campanella and Schipano as based partly on their interest in creating an astrological-medical horoscope of Galileo. Accordingly, Campanella wrote to Galileo in early 1614 asking for his medical history so as to compose a medical horoscope. Galileo rejected the offer. Later that year, Amabile tells us, Fabio Colonna and Mario Schipano engaged in the same practice concerning Federico Cesi's deceased son; Cesi seemingly believed in medical astrology. By looking closely at the son's medical history, Schipano and Colonna deduced cosmological causes for the tragic death. Consequently, Campanella wrote once again to Galileo, but now using the story of Schipano and Cesi to convince him to send his medical history.[87]

In the end, Giovanni Bellori, who wrote della Valle's *Vita*, supplies us with decisive evidence that della Valle was a follower of Campanella. He writes that on della Valle's return to Rome, the Romana Accademia degli umoristi held a reception to honor him as one of its most successful alumni. Later, della Valle opened a small academy at his home, a circle of literati who studied the theology of Campanella.[88] This small academy also included a kind of cabinet of curiosities, where della Valle displayed the many objects he had collected in the Near East.

In tracing the social and intellectual threads of Pietro della Valle's travel from Goa back to Naples, we discover his personal connections with actual men—contacts made through friendships and common agendas, not just by subscribing to their ideas. Moreover, the cluster of Galileans in Naples was more than an intellectual source; it included the patrons for whom della Valle labored on his travels to collect ancient manuscripts of Holy Scriptures, medicine, and natural philosophy.

During the 1610s, Mario Schipano, a Neapolitan natural philosopher and historian and a senior member in secretive Pythagorean circles, was involved with the Lincei Society and the Galilean affair. The question of the compatibility of heliocentrism with the Holy Scriptures preoccupied these circles. In the Neapolitan context, the Galilean affair involved Galileo, Foscarini, della Porta, Colonna, Campanella, and Schipano, acting in a personal network. In 1614–15, at the peak of the intra-Church struggle over Copernican cosmology, Schipano was in close touch with Tommaso Campanella, who wrote in prison an exceptional apologia on Pythagoras, Copernicus, and Galileo, arguing for the existence of Jewish sources of Pythagorean heliocentrism in the Holy Land. At the same time, Schipano sent Pietro della Valle, a young, educated Italian who had some experience with the Islamic world, to the Levant to collect manuscripts in Arabic, natural history, and ancient Holy Scriptures. After the first trial of Galileo (1616) and inclusion of the Hebraist de Zuñiga, Copernicus, and Foscarini in the Church's index of prohibited books, as well as rejection of the de Zuñiga–Foscarini compromise between the Copernican system and Scripture, Schipano and others shifted the debate from "proper interpretation" of Scripture to "authenticity of the Scriptures." Using the reformist Hebraist critique of the Vulgate, they started looking for the most ancient and original versions of the Bible in Hebrew and Chaldean, to show that ancient knowledge of the secrets of nature, or the Pythagorean schemata of nature, had, like Copernican cosmology, fallen between the translational cracks of the scriptural versions (the Mesora Hebrew, the Septuagint Greek, and the Vulgate Latin Bibles).

With his agenda set, della Valle first studied Arabic, Persian, and Hebrew for a year in Constantinople, where he began collecting manuscripts. From this point onward, his itinerary modeled that of Pythagoras, and in Egypt he collected evidence such as mummies and looked for the Holy Scriptures of the Samaritans, which chronologically preceded all other available versions. Later he traveled to Sinai and looked for Mount Sinai, where Moses first wrote down the Scriptures. In the Holy Land and Syria, della Valle finally obtained Samaritan scriptures. Some time after 1616, Schipano, who followed the tradition that the Book of Job was originally written in Chaldean, informed della Valle about the Galilean affair and the inquisitorial decree, and instructed him to look for the "original" Book of Job. Hence, on his arrival in Mesopotamia, the alleged land of Job, della Valle started looking for the Book of Job in the Chaldean language among local groups whose lineages traced back to the ancient Chaldeans and among local Jewish communities that may have been there since the First Temple and kept pre-Mesora versions of the Bible. In Lār, he found a sparkling intellectual culture with links to the ancients and was excited by the work of a local astronomer, Zayyn al-Dīn al-Lārī, whom he tried to convince to convert to Catholicism. In Lār, he also looked among the local Jewish community for a Chaldean Book of Job. From there he continued to travel among the Christian Chaldeans, and in 1623 he arrived in Goa, where he met the Jesuit Christopher Borrus and translated the latter's work on the Tychonic system for al-Lārī. He finished the translation, probably in the 1630s, in Rome. The letter was to be sent to al-Lārī for two purposes: to further convince him of the superiority of European astronomy and consequently the Christian religion, and to instruct him that he should not entirely accept the Tychonic system. Della Valle stressed that if al-Lārī would continue the search among the Jews of Lār for the "original" Book of Job in the Chaldean language, then finding the "original" version of Job 9:6 would solve the Copernican-Tychonic controversy.

The encounter between scriptural commentary and Copernican cosmology, which developed from the tactic of accommodation to the question of the original Scriptures, extended the Galilean affair into the ancient Near East. The hermeneutic dynamism found in an array of strategies and tools necessitates a reevaluation of our usual narrative of the Galilean affair as a purely local Christian theological debate. Instead, the reconciliation of Copernicanism with Holy

Scripture involved a hermeneutic leap from accommodationism to radical Hebraism. It generated travels for the purpose of collecting ancient manuscripts of "lost scriptures." The ur-text of the Book of Job acted as a historical a priori, or ancient authority of truth, that could resolve cosmological controversies and reconcile Pythagorean cosmology with Scriptures.

Transcending Time in the Scribal East

IN 1629, JOSEPH SOLOMON DELMEDIGO published in Amsterdam a somewhat incoherent book entitled *Sefer Elim* (Book of Elim). The book was printed by Menasseh Ben-Israel (Spinoza's teacher) and was a collection of articles on natural philosophy and mathematics. Before his arrival in Amsterdam, Delmedigo had traveled in the Eastern Mediterranean (1616–19), where he participated in a public contest in mathematics, collected ancient manuscripts, studied the Cabala, conducted observations of the 1619 comet, and was associated with the Karaites, a Jewish sect that rejected rabbinical literature in favor of the purity of Scripture. In *Elim*, Delmedigo promoted the Copernican cosmology to which he was exposed by Galileo, his teacher at Padua University during the years 1606–13. The book gives only vague clues for connecting Delmedigo's Copernican conviction with his travels. However, a close reading of this and other published works, set alongside his culture of learning and personal associations, helps us understand the connection he made between advocating Copernican cosmology and collecting ancient manuscripts in the Near East.

On his arrival in Amsterdam in 1629, Delmedigo met Menasseh Ben-Israel and showed him the collection of manuscripts he had brought with him from the East. Delmedigo suggested publishing some Karaite manuscripts that showed how some of the astronomical technicalities associated with the new post-Copernican astronomy were already known in the ancient past. Rejecting that idea, Ben-Israel recruited Delmedigo to write a book in Hebrew about the new natural philosophy. While *Elim* ostensibly presents current questions in the fields of natural philosophy and astronomy to readers of Hebrew, a close reading of the title, the text, and some suspicious inconsistencies within it reveals that Delmedigo had a much more ambitious goal in mind.[1]

The few historical investigations of Delmedigo have considered his work as promoting Jewish interest in the "Scientific Revolution." But most emphasis has been given to the inconsistencies in the text of *Elim* and the incomprehensibility of Delmedigo's writing. Abraham Geiger, who in the nineteenth century attempted to reform Judaism and accommodate it to the changing times, saw Delmedigo as the precursor of the eighteenth-century Haskala (Jewish enlightenment) movement in Germany, which integrated rationalism and modern philosophy into Jewish intellectual culture. Geiger was the first to consider the two seemingly contradictory motivations of Delmedigo: the new astronomy and the Cabala. For Geiger, Delmedigo showed interest in the Cabala merely to pay homage to his patrons, while at heart he was a rationalist.[2] Geiger led a generation of historiography about Delmedigo, giving us these two poles: Cabala and rationalism.

Since the 1970s, a new interest in Delmedigo has surfaced. Isaac Barzilay, for instance, sees Delmedigo's "inconsistencies" between rationalism and mysticism as reflecting the polymathy of the time, when scholars sought evidence and experience through travel and spread knowledge with no particular stance. "His real interest," Barzilay argues, "lay in the pursuit of those studies to which he was introduced in Padua."[3] Joseph Levi suggests that Delmedigo was "influenced" by the scientific societies in Italy, especially the Lincei Society—of which his teacher, Galileo, was the seventh member—and that Delmedigo struggled to establish a Jewish scientific society.[4] Finally, David Ruderman dismisses the seeming inconsistency between reason and mysticism in Delmedigo's work, especially in the light of the new historiography of science pioneered by the work of Frances Yates.[5] Ruderman simply gives a direction to follow, and here I pursue it further by looking into specific peculiar aspects of Delmedigo's writing.

We can also look at recent studies of skepticism, which have regarded figures such as Delmedigo as part of a wider trend that sloughed off the assumptive bases of biblical scripture and Aristotelian philosophy. Joseph Kaplan, for instance, presents Delmedigo's interest in the Karaites as part of a larger interest, especially on the part of Protestants, who held that the Karaites were the "true Jews" (or "Protestant Jews") who dismissed oral tradition and relied solely on the Bible.[6] The late Richard Popkin linked this observation to apocalyptic sentiments that flooded Europe after the Reformation, the discovery of the New World, and the changing cosmology, all of which helped foster the idea that the Karaites, at the end-time, would lead the return of the lost Ten Tribes to the Holy Land.[7]

None of these accounts tries to understand Delmedigo's "inconsistencies" in the light of his readings in cutting-edge natural philosophy and his travels in search of ancient manuscripts and other knowledge. Yet it is important to pay attention to Jewish natural philosophers and to understand their motives and frustrations in the face of the "Scientific Revolution" and the print culture. Thus, the intersection between previous works on Delmedigo and the new studies of skepticism could be combined with approaches taken in the history of science and the history of the book. In this chapter, we look at print artifacts, such as approbations for Delmedigo's work, comments by printers, and examples of self-censorship, as well as the problems in transforming manuscripts into books.

A key question concerns Delmedigo's travel to the Eastern Mediterranean. How might his Copernican convictions have motivated his travels? In cross-cultural exchanges and literary rhetoric, he fashioned himself as a foremost student of Galileo and as a leading Jewish representative in natural philosophy. He traveled to the Eastern Mediterranean to restore ancient Jewish knowledge about nature, and he did so by searching for and collecting manuscripts of ancient wisdom, an alleged *corpus hebreicum* (to coin a phrase) that had escaped the censorship of the rabbinical printing presses. Delmedigo traveled, collected objects, and observed celestial events against the backdrop of a grand quest for the mythical Elim, a utopian biblical locale where, according to cabalists, the revelation of the secrets of nature had commenced and had spread to the whole universe. He linked Jewish-hermetic revelation with Copernican cosmology and sought material objects such as ancient Hebrew manuscripts that, purportedly, maintained a stronger connection to the revelation at Elim. Subtle hermeneutical trends of his day stressed that the pure knowledge of nature, including the discoveries of Copernicus and Galileo, had been given to Moses and the Jews in deepest antiquity. The Jews, Delmedigo argued, eventually lost their leading role in natural philosophy to the Gentiles because of stagnation and narrow-mindedness in the rabbinical intellectual culture. Delmedigo cultivated the myth of Elim to provide an inclusive account of the Copernican cosmology that encompassed Jews. The so-called Copernican Revolution actually could be seen, if demonstrated properly with ancient manuscripts, as being linked to ancient Jewish theology.

Practices and Associations in Cairo and Constantinople

What can we learn about Delmedigo's cultural, social, and intellectual practices in the Eastern Mediterranean? All of his works were printed in Europe, mainly in Amsterdam, and due to the printing world's cultural containment and self-censorship, he felt it necessary to obscure his motives for publication.

Figure 16. Portrait of Joseph Solomon Delmedigo (1591–1655) and the frontispiece of *Elim*. From Delmedigo, *Sefer Elim* (1629), Heb 7415.75. By permission of Houghton Library, Harvard University.

Elim (figure 16) gives some details about the years Delmedigo spent in Cairo and Constantinople (1616–19), but is only sketchy about encounters and exchanges with local scholars. However, his manuscript entitled "An Article on a Comet" mentions his observations in Constantinople on the comet of 1619 and includes anecdotes that give a sense of how he introduced local scholars to the Tychonic explanation for the comet. At the beginning of the manuscript he writes, in rather inaccessible handwriting:

> In 1619 in Constantinople, several colleagues, rabbis, and Karaites who used to listen to my voice and study with me, woke me up in the middle of the night and said: "Why are you fast asleep, while you have wisdom of the heavens and understand the signs from above?" Then they took me out and said, "Look at the heavens." I raised my eyes and saw a wide and long-tailed star [comet] burning and shining in the sky. I had never seen such a comet, and I had not read others' works on something as big as this.[8]

He goes on to describe how he disabused the locals of any occultist interpretations.

> Some of the crowd gathered around me, expecting to hear my teaching. Since they were terrified of this great novelty [the comet], they said: "Not for nothing has God sent this star, but instead to be a sign and precursor for a forthcoming event." But I denied their requests [to make an astrological prediction] and told some of them that it [the comet] is a natural thing . . . I was a boy and their thoughts were not mine and my ways were not theirs. I was interested to find the comet's location and distance by using mathematics and trigonometry [parallax], and for this purpose I observed it for a few nights with a quadrant from a high place in the city. From these observations and calculations, it became clear to me, and others, who were there with me and who agreed or disagreed with me, that it is impossible that the comet could be in the world of the elements [the sub-lunar world], but it has to be in the heavens in the ethereal world.[9]

Delmedigo refers to Copernicus and Tycho to confirm his observations and calculations, and he ends the short manuscript by refuting, in an interesting alchemical way, the Aristotelian explanation that the comet could be a gaseous phenomenon in the sub-lunar world. After calculating how much of the earth's water would be needed to create a celestial phenomenon or a gaseous body "a thousand times bigger than Earth," he concludes that the comet could not be a gaseous body. At the end, Delmedigo contextualizes the significance of his findings for what he describes as the Jewish-Christian "contest" in natural philosophy. The purpose of his writings and activity, he says, is "to show them [the Gentiles] our strength, and they should know that the Children of Israel are not the lightest in this race on the difficult issues of natural philosophy. So, I would like to announce by demonstration to everybody that the noticed comet was more than 100,000 times bigger than the body of Earth. It should be clear to all the respectable astronomers and observers in Germany, Holland, and Italy that [the comet] was part of the sub-lunar world."[10]

Observing and objectifying the comet allowed Delmedigo not only to establish astronomical arguments but to create cultural capital. As a Jewish astronomer, he had something new to contribute to the astronomers of Europe who observed the comet but were consumed with Aristotelian explanations. Delmedigo keenly called for an end to Jewish intellectual isolation. He made sure to mention that he studied techniques of observation and calculation, especially celestial distances and light magnitudes, from non-Jewish professors in Padua. In a discussion of how the intensity of planetary light depends on the distance between the sun and Earth, he writes: "my teacher Galileo, when he researched Mars, showed that when it is close to Earth its light surpassed the light of Jupiter, although its body is much smaller. Moreover, its light is so great that the eye almost cannot look at it through the telescope [*sheforferet*]. When I asked to observe it with the glass instrument [telescope], Mars was long and not circled [by anything], probably because of the motion of its sparks of light, and when I looked at Jupiter and Saturn [through Galileo's telescope], they had a circular form like an egg."[11]

In another place in *Elim*, Delmedigo once again mentions his affiliation with Galileo, to establish personal credibility—"The images of the stars change as much as they are close to [i.e., the closer they are to] the sun, as I observed several times through the glass of Galileo"[12]—and thus gives us the only evidence of students of Galileo who observed celestial objects through his telescope. To be sure, Delmedigo's call for revival of Jewish natural philosophy, combined with his admiration of Galileo, would soon characterize a particular take on the new cosmologies. While stressing the findings of Galileo, Delmedigo was disturbed that no Jew had a role in these great innovations and no Jew other than himself was intellectually able to appropriate and understand them. He fashioned for himself a scientific persona of a Jewish representative in competition with Gentiles over natural philosophy.

"The Boy" Went Down to Egypt

There are specific clues in Delmedigo's anecdotes that help us understand how his scientific persona developed. In "Article on a Comet," he describes himself in Constantinople as "a boy," even though he was about twenty-eight and already experienced in observing celestial phenomena and debating publicly with Middle Eastern natural philosophers and mathematicians. At this stage he had been traveling for three years since leaving his home on Crete (1616), and six years since graduating from Padua University (1613). Before Constantinople, his trip had taken him to Cairo for just under a year, between 1616 and 1617, during which time Pietro della Valle was there, looking for the Samaritan Pentateuch. In Cairo, Delmedigo debated mathematics and natural philosophy with Jews, Karaites, and Muslim scholars. We know little about his stay but can deduce two important personal developments.

The first is that he made a lifetime friend in a Karaite scholar named Jacob Iskandrani.[13] *Elim* refers to the latter's commentary on Euclid,[14] and stresses that Delmedigo appreciated Iskandrani mainly for his intellectual capacities and masterly understanding of Euclid's *Elements*. In another work, *Melo hofanim*, Delmedigo mentions him as one of his sources: "In the city of Cairo in Egypt I saw the elder noble Jacob Iskandrani, wise and knowledgeable in scholarship, who added his own contribution to mathematics in his commentary on Euclid's *Elements*."[15] Moreover, he dedicated another of his essays—*The Wonders of God* (*Niflaot haShem*), which is now lost but is mentioned in *Elim*—to Iskandrani.[16] His attachment to Iskandrani not only was based on personal friendship, but also resonated with Delmedigo's concern over rabbinical intellectual culture. While he kept arguing that no Jewish rabbi was a worthy colleague in natural philosophy, Delmedigo expressed high appreciation for the Keraite Iskandrani, considering him one of only three Jews (the others were Simone Luzzato and the Keraite Zerah ben Natan) who could grasp the new mathematics and astronomy.[17]

A second theme emerging from Cairo is connected to the first. While stressing the lack of Jewish partners in natural philosophy, Delmedigo highlights his role in saving the Jews from humiliation in that field. In Cairo he mainly associated with Karaites, but as a by-product of this connection, he was also introduced and exposed to Muslim scholars. One of them was an Egyptian mathematician whose name Delmedigo corrupted as 'Ali Ben Rahmadan, but probably should be 'Alī Ben Raḥīm al-Dīn, a professor of mathematics in one of the local colleges (*madrasahs*).[18] This encounter led to one of the most interesting of Delmedigo's anecdotes. He introduced himself to the Muslim scholars as a mathematician and student of the great mathematician Galileo, but the name of Galileo was not known in the intellectual circles of Cairo, where Europeans were still seen as inferior in natural philosophy and mathematics. The introduction intrigued 'Alī Ben Raḥīm al-Dīn, who challenged Delmedigo in an attempt to demonstrate Muslims' superiority in mathematics. Here is the challenge, as given in *Elim*: "'Ali Ben Rahmadan the Egyptian used to teach in the academies of the big city of Cairo. The man was senior among the Muslims and he was arrogant. When I came there he tested me in riddles and paradoxes. I used to avoid him and hide, since I lacked books of reference." Delmedigo contrasts their scientific personas—a traditional old scholar versus a young vivacious scholar.

> He was quiet and peaceful in his house, like wine in an old bottle, while I was poured from one vessel to the other, wandering, stranded in exile. He urged me [to contest him in mathematics] many times until I was embarrassed and had to see him face to face. We met one day and transmitted, one to the other, ten questions and we fixed a time in four weeks for the confrontation. At the agreed day and place, I found a crowd of Muslims, Jews, Karaites, and Samaritans all

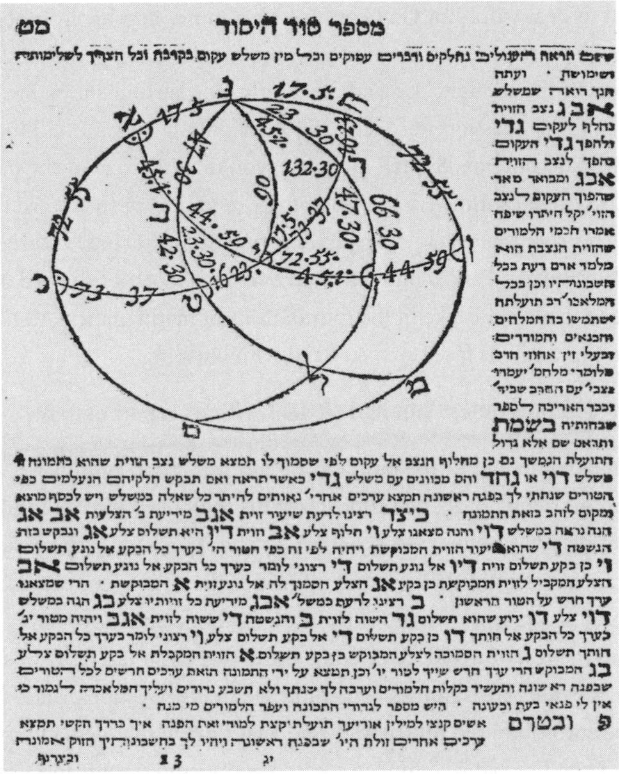

Figure 17. Spherical trigonometry in *Elim*. From Joseph Solomon Delmedigo, *Sefer Elim* (1629), Heb 7415.75. By permission of Houghton Library, Harvard University.

> gathered there to see whose truth would stand. I was disappointed that the jury decided that I would be the first to answer all of his questions, and deliver them in a neat handwriting, since I was a boy and he was an old man experienced in many ventures. What I thought would be a shortcoming turned to be a salvation for me.

The Egyptian mathematician presented a question in spherical trigonometry (an area of mathematics later included in *Elim*; see figure 17), asking Delmedigo to find the size of the parts of a spherical triangle, not knowing that the sum of the angles should be larger than the angles of a perpendicular triangle. Delmedigo was not sure about the intent of the question.

> His triangle, therefore, was false and I thought he was trying to deceive me and test me with a tricky question. I said that it is a false question that I have no

> intention to deal with. But God confused his mind, and he did not admit to [trying to deceive and] say "I tested you and aimed to see if you would detect the defects of the triangle." Instead, he laughed at me and shook his head, affirming that the question was valid and it was my mistake. I told him, if this is true, why do not you, Sir, tell me what would be the size of the part of the triangle. In the meantime I asked one of the members of the crowd to go and bring the book of Menelaus. After he ['Alī] finished lecturing for three hours, I opened the book and asked him if he had ever read it. And I showed the crowd that he had made a mistake in the foundations of mathematics. All the people stared at him until his face [was] covered with blushing.

Delmedigo won the contest, but had to deal with its repercussions.

> It was already the end of the day, and he left upset and humiliated and shut himself in his house, so that he almost lost his soul. I was terrified and shaking from [fear of] the Muslim crowd that might have targeted me, because, for our sins, we are subordinated to them. And one of the elders sent me, escorted by several men, back to my home. Since then my fame and glory have spread in Egypt and the rest of world. But do not think that this man ['Alī] was empty, I swear he was full of wisdom as you could see in his questions. Later I went to his house and submitted myself and said that I knew that he had just made a mistake. Then I recognized him as a wise and ingenious man and I said that I would like to learn from him just like one of his students. So he hugged me and respected me in front of the elders of his people.[19]

The skewed dialogue between the representative of the new astronomy and mathematics (Delmedigo "the boy")[20] and the traditional Muslim natural philosopher ('Alī Ben Raḥīm al-Dīn, "the old man") was a microcosm of the first encounters between European and Near Eastern natural philosophers and mathematicians after the Copernican Revolution. Mathematics was becoming a tool to judge supremacy in a cultural-scientific contest between Europe and the Near East. Lacking significant information about 'Alī Ben Raḥīm al-Dīn, we can only speculate that he at least was suspicious of the new natural philosophy and mathematics. Nor is there anything to tell us how either the Keraite Jacob Iskandrani or members of the crowd received the new astronomy brought by Delmedigo or the news of the contest. With such gaps in our story, we must use other clues to get to Delmedigo's world of imagination and mythology.

Collecting Manuscripts and Criticizing Print Culture

On his trips, Delmedigo collected, read, and developed an enormous thirst for unknown (to Europeans) manuscripts. In a short biography of Delmedigo, Moshe Metz, his student, described him as an omnivorous reader: "Delmedigo swallowed in his stomach many books, and never spared either his money or possible burden and far distance in order to collect books from whatever [place] he saw, heard of, or even was aware of. His treasure amounted to 7,000 books, with a price of 10,000 gold coins . . . We cannot know whether in the whole universe there is anyone who surpasses him in the searching and striving for books."[21] Delmedigo cultivated this image of a reader and a collector of books, intellectual practices that came at the expense of his private life—as he put it: "I was unfortunate with sons and property, but my luck was great with books, and there was no precious book in the world that I desired and could not obtain, and sometimes I wandered hundreds of miles by land and sea in order to see a small book, even though occasionally I was disappointed."[22]

Yet Delmedigo was not merely a collector but a scholar who looked for diverse sources for his intellectual projects. He mastered classical sources and claimed to know by heart "a few books of ancient authors like Euclid, Menelaus, Ptolemy, Archimedes, Aristotle, Plato, Galen, and Hipparchus."[23] Beyond the classical corpus, Delmedigo believed in the existence of a *corpus hebreicum* that could show how the notion of heliocentrism had an ancient Jewish lineage. This combination of a scholar fluent in ancient sources and a collector searching for manuscripts in the Eastern Mediterranean emerged from a growing critique of radical Jewish scholars from Venice and Amsterdam, who perceived the intellectual culture in early modern Europe as narrow-minded.

The practice of collecting manuscripts implies that with the rise of print culture, certain knowledge and sources from the past would be lost. In the introduction to Delmedigo's *Novellas of Wisdom* (*Novlot ḥokhmah*), a book on the Cabala, we find that Delmedigo went to Egypt to look for manuscripts of Karaite exegesis and Jewish natural philosophy that never came into print. The poverty of print culture prompted him to conclude that "we [Jews] lost our knowledge and our wisdom."[24] Printing had fixed the religious priorities of rabbinical culture and its canonization of Jewish works, while excluding nonreligious works from "Jewish library shelves." Whatever was not printed was doomed to be lost.

Delmedigo had many reasons to complain. An examination of the lists of printed books of the major presses in Italy during the sixteenth and early seventeenth centuries indicates that printers excluded works on philosophy and natural philosophy. Most printings of Jewish/Hebrew books were concerned with Holy Scriptures, Talmud, and liturgy. Not even the philosophical works of medieval Jewish philosophers and astronomers, such as Maimonides and Ibn-'Ezra, were included.[25] The printing revolution refashioned sources of Jewish learning. Instead of the dissemination of scientific texts and the flourishing of science, print technology led to the disappearance of natural philosophy texts from Jewish intellectual life. Hebrew libraries in the academies became useful merely for finding books on oral law. Delmedigo blamed rabbinical narrow-mindedness that cast "a darkness that covered the earth and now many are ignorant."

> Even though the land is filled with Jewish academies the quality of teaching and research is poor. The many people who claim to be scholars are only looking for a living and for material fulfillment. Even those who are great scholars in the Jewish law, their aim is entirely to be teachers or judges or heads of academies, and this is an evil disease . . . and eventually they end up with no understanding of natural philosophy and as a result with misunderstanding of the Torah as well. They become enemies of wisdom. If someone should be interested in mathematics, astronomy, or Greek philosophy they would see him as one who sleeps with a Gentile woman.[26]

The economic-cultural priorities of Jewish printers constituted a certain pragmatic censorship, which in turn created a break from ancient Jewish natural knowledge. The curricula of Jewish academies also reflected a type of censorship—namely, the exclusion of medieval and ancient writings on natural philosophy. In the Jewish intellectual centers of Europe, such as Venice and Amsterdam, the available sources were limited to printed books; scholars ceased to engage with the Jewish-authored manuscript sources of natural philosophy. For Delmedigo, print culture posed a great impediment to Jewish intellectual activity. The Near East, which lacked a print culture, became in a sense a "mythical field of sources" where Delmedigo could look for the mythical *corpus hebreicum*. The scribal Eastern Mediterranean offered a freer and more open-minded culture.

The quest for the *corpus hebreicum* dictated Delmedigo's itinerary between 1616 and 1622, which included Egypt, Constantinople, Crimea, and Poland, locales that had significant Karaite communities. *Elim* tells us that Delmedigo "first traveled to Egypt, where he found with great joy many new and old books of the wise men of Africa, which he had never seen before."[27] Then he went on by sea to Constantinople, skipping the route through the Holy Land, where there were no Karaites.[28] He went directly to Constantinople because "whoever wants to see precious and rare books should go to Constantinople, the mother and father of scholarship."[29] Delmedigo, ostensibly, believed that the Karaites and their ancient manuscripts had pure lineages to the ancient revelation.

From Lost Karaite Texts to Karaite Thought

The attraction to the Karaites might be explained by their interest in contemporary natural philosophy and their struggle with rabbinical intellectual culture.[30] In addition, the Gentiles perceived the Karaites as potentially better able to be integrated into Gentile intellectual culture than were Jews in general.

We have only vague information on the origins of the Karaites. Their own accounts trace back to controversies at the time of the First Temple, but historical accounts argue that the sect started in the eighth century. The founder, ʻAnan, challenged the authority of the oral Jewish tradition (Midrash, Mishnah, and Talmud) and stressed sole reliance on the Scriptures and a literal understanding of them through reason.

In the tenth century, the Karaites' critique of tradition brought them close to the philosophical circles of the Mu'tazilah, a philosophical school that flourished in eighth- and ninth-century Baghdad and Basra, which advocated rationalism (at the expense of tradition), free will, and atomism. The tenth-century Muslim historian al-Mas'ūdī referred to Karaites as "Jewish Mu'tazilites."[31] Karaites' association with the Mu'tazilah also brought about some Karaite works on atomism.[32] In philosophical discussions they held that "causality" is a better notion than "will" to explain the world, because the fundamental principles of physics are the atoms and void that cause motion.[33]

Such Karaite philosophers as Yusuf al-Basrī in the eleventh century prioritized rational ethics over tradition.[34] His pupil Jeshua ben-Jehudah held that knowledge of the creation cannot be derived from Scripture alone and is subject also to rational speculation.[35] Later, Aaron Ben-Joseph (Ahron Ben-Josef) the Physician (1250–1320), of Constantinople, modified Karaite natural philosophy by subscribing to the Aristotelianism of Maimonides. He maintained that the world is made up of four basic elements and not of atoms.[36] However, the short flirtation with Aristotelianism ended several decades later, when Aaron ben Elijah (1317–69) renewed Karaite atomism in his work *Tree of Life* (*Etz hayyim*), which aimed to be the counterpart to Maimonides' *Guide for the Perplexed*. Ben Elijah's book suggested that the world was created from and maintains natural laws concerning its constituent parts—atoms. Creation, for ben Elijah, meant a combination of atoms, and dissolution of the combination would mean the separation of atoms and thus changes in nature.[37]

Delmedigo, then, had grounds for engagement with the Karaites, especially atomism and critical Aristotelianism, and their approach served him well: his search, after all, was for ancient evidence for recent findings and discoveries in natural philosophy. The outward-reaching attitude of Karaites appealed to his call to reinvigorate Jewish intellectual culture.

The renowned eleventh-century astronomer and exegete Avraham Ibn-'Ezra was also a link between ancient theology, Karaite writings, astronomy, and Jewish prestige. He was a role model of rabbinical intellectualism, especially through his amalgamation of natural philosophy, astronomy, and exegesis. Delmedigo refers to him as "a great wise man, the leader of literal understandings of the text,"[38] and adheres to Ibn-'Ezra's hermeneutic that directly addressed the inconsistencies, misspellings, and exaggerations in the Holy Scriptures. Flaws were clues to the secrets of creation, deliberately embedded in the text.[39] Ibn-'Ezra's surrounding intellectual culture, especially of Karaites, fashioned his philosophical style.[40]

Major early modern thinkers highlighted Ibn-'Ezra as an important source.[41] Robert Westman even suggests that the title of Copernicus's major work, *De revolutionibus*, reflects the wording of Ibn-'Ezra's *Liber de revolutionibus et nativitatibus*.[42] The growing interest in Ibn-'Ezra resulted in an increase in printings of his works in Latin as early as the sixteenth century.[43] Printings mostly concerned Ibn-'Ezra's astronomy and astrology. However, such texts were not among the priorities for Hebrew presses,[44] and thus Ibn-'Ezra's manuscripts in Hebrew remained available in the Hebrew scribal culture in the Eastern Mediterranean.

Delmedigo made Ibn-'Ezra a Jewish scientific icon and aspired to continue his work to promulgate ancient Jewish astronomy and timekeeping to the Gentiles. Emulating his icon, Delmedigo acted as a personal "center of calculation" that travels, collects, and reworks various extant sources on astronomy. From Salonika he obtained Moshe Almosnino's Hebrew translation of Peurbach's *Theoricae novae planetarum* and critically discussed it, especially in the light of the new discoveries in astronomy that rejected the existence of crystalline spheres.[45] From Istanbul he acquired Arabic astronomical works from the fifteenth-century observatory of Samarqand and worked to translate the updated astronomical tables.[46] From Greek scholars in Istanbul he got hold of Greek versions of Ptolemy's *Almagest* and compared them with Arabic commentaries on Ptolemy.[47]

As one of only a few Jewish scholars cited in the literature of post-Copernican astronomy, Ibn-'Ezra served Delmedigo's grand design as a link between the ancient Jewish wisdom and the new Copernican Revolution. The manuscripts of the Karaites and Ibn-'Ezra were threads to the alleged *corpus hebreicum*. Their exegeses were direct, non-allegorical readings of the Scriptures, and as such they avoided the filtering that could alter ancient, perfect knowledge. Writings of Karaites and Ibn-ʿEzra combined exegeses with deep commitment to the exploration of natural philosophy. Delmedigo thus promoted Copernican cosmology as something prefigured in ancient theology.

Hermetic-Cabalist Critique and the Role of Ancient Jewish Knowledge in a Post-Copernican World

The practice of collecting ancient objects, specifically manuscripts, was posed against the myth of Elim—a Jewish-hermetic revelation of natural philosophy—and its textual remnant, the *corpus hebreicum*. According to Delmedigo, the study of nature does not contradict Jewish law but, rather, strengthens the study of the Torah, because natural philosophy is the study of the deeds of God in his most sensible, tangible, and graspable extension—nature.[48] Delmedigo's hermetic-cabalist reading of the history of natural philosophy, however, went even further. It affirmed that a perfect knowledge of nature was revealed to Moses, but was then transmitted along two parallel channels: from Moses to the prophets and to Gentile philosophers such as Hermes and Pythagoras. In this way, Delmedigo formed a meta-narrative in which he criticized Aristotelian philosophy for having lost the connection to its ancient sources and presented post-Copernican cosmology as having regained it.

The rabbinical intellectual culture failed to incorporate the new cosmology or to understand its Jewish origins. Delmedigo underscored in *Elim* the ancient Jewish lineages of the knowledge of heliocentrism that was "in ancient generations ours, and from the stomach of Judea it came out."[49] In ancient times, the Jews had mastered the knowledge of nature before it was swept up by Aristotelian philosophy. The central aim of Delmedigo's search, then, was to recover the historical origins of man's knowledge about nature and its spread across different cultures and down to his own time. In the introduction to *Elim*, Delmedigo made an emotional complaint that philosophy was stolen from the Jews: "The wisdom was given to Moses in Sinai, and it was carried by the prophets from Jerusalem. The Greeks of Athena searched in each one of its corners to steal wise things. There is no justice; they received joy and we the misery, because it [the wisdom] left us."[50] Elsewhere in *Elim*, Delmedigo states that the "highest revelation about the creation and the structure of the universe was the one that was given to Moses."[51] The revelation to Moses turned out to be the source of all philosophies: "The sages of the Talmud used to say that the wise Greeks came from Athens to Jerusalem to listen to lectures. At first, Israel had all wisdom and reason. It was given to Moses in Sinai and transmitted to the prophets, and most of the natural philosophers, poets, and physicians were Hebrew. The wisdom of Cabala was also first given to the elders of Jerusalem. However today, for our many sins, our mind has weakened and our wisdom rotted. No one is left [among the Jews] that is knowledgeable in the art of astronomy, natural philosophy, and other arts external to the Oral Law."[52]

Delmedigo considered it his vocation to rectify this historical accident by collecting manuscripts that represented a pristine revelation that included actual textual remnants, thus cutting through the "filtering" by the printer-rabbis and recovering Jewish glory in natural philosophy. The mysterious place and time of the pristine revelation are evident in the title he picked for his book.

By instinct, we tend to think that the title *Elim* could be a literal translation of a famous seventeenth-century alchemical book, *Mutus liber*, a mute and wordless book—a collection of images and illustrations of alchemical practices, with no written text. *Mutus liber* was published in 1677, however, some years after Delmedigo died. But there was a long tradition of so-called mute books in alchemical writings. The connection made between Delmedigo's *Elim* and *Mutus liber* implies that the interest in alchemy included in *Elim* somehow echoes a current genre of mute books.[53] Besides the alchemical implications, the title also points to aspects of Jewish tradition.

The word *elim* does not have a fixed significance in Hebrew, but we may speculate. Delmedigo explains that the structure of his book, which consists of twelve "articles" and seventy "doubts," was modeled on "the twelve springs and the seventy palms."[54] The reference is to a biblical verse, Exodus 15:27: when the Israelites were exhausted en route from Egypt, one of their first stops was at Elim, a utopian oasis in the desert "where there were twelve springs of water, and seventy palm trees, and they encamped there by the water." Some medieval commentaries say that in the creation of the world, God created the twelve springs of water to stand for the twelve tribes and the seventy trees of palms for the seventy wise men of Israel.[55] Others, including Ibn-ʿEzra, held that the place represented wealth given not only to the sons of Israel, but to all the world. Cabalist interpretations suggested that "twelve springs" represented the twelve channels of spiritual wealth that came down to the world, and "seventy" was the seventy sacred names of God with which one can decode the secrets of nature.[56] Delmedigo followed earlier usages of a similar title and structure for a book, as used, for example, by the early sixteenth-century cabalist Moses Cordovero, who wrote a book called *Elima* ("going toward Elim") with the same structure.[57]

Gershom Scholem's marginalia on the Zohar, a major cabalistic commentary on the Pentateuch, can help us further. In the relevant Zohar passages, Scholem notes that the sources were referring to the generation that arrived at Elim; this generation was called Darda (*dor de'ah*), "the generation of science," to give the Aramaic word its true weight. Members of the new civilization explored the present—natural phenomena in the here and now. They set aside metaphysics and theology and explored the "book of nature." In Elim, the process of revelation started with a knowledge of nature, and only later culminated at Mount Sinai, where theology, ethics, and metaphysics came into play. Scholem's sources also linked the verse concerning Elim and the generation of knowledge to King Solomon. Just like the Elim generation, Solomon, who was wiser than the ancients, had explored nature, talked to animals, and eventually encompassed all knowledge, including the "secular."[58]

The cabalistic view, of course, heavily included numerology: the numerical equivalent of the word *elim* (through letter-numerical symbolism) in Hebrew is the sum of seventy plus twelve.[59] Assuming that the sources used by Scholem were also available to Delmedigo, we can infer that he used a coded language and addressed the title of his book to readers of esoterica. To show that the new cosmology was not completely new, Delmedigo could feature his myth of Elim, a historical and geographic locale where the oldest and purist revelation of natural philosophy was given to Moses.

Delmedigo's hermetic views went beyond the question of the revelation to Moses and touched on practices of alchemy and magic. Zerah ben Natan, a Karaite scholar, wrote to Delmedigo that he had heard that "you [Delmedigo] preach to every Jew that this [practice of hermeticism] is wisdom, and you master all of its aspects and details, and your student even showed us some of your spectacles, like various crystals you made, and the firm iron you made out of the soft one, and the transformation of metals to silver, and an oil that would rejuvenate an aging face."[60] In fact, at the end of *Elim*, Delmedigo discusses hermetic practices to reject the Aristotelian central reliance on experience and metaphysics. He connects practices of hermeticism to experimental science and declares that "I do not approve as truth anything except what I have experienced and examined several times. Whatever I have not examined or we did not happen to see, I would neither approve nor deny."[61]

Returning to Delmedigo's hermetic-cabalist natural philosophy, we must make the connection to post-Copernican cosmology. What examples of lost ancient Jewish knowledge were invoked by post-Copernican astronomers? Astronomy before Tycho Brahe conceived the motion of planets as a motion of crystalline spheres in which the bodies were embedded. Delmedigo presents various schools and their views to show that the only opinion compatible with Tycho's was that of the Jewish sages of the Second Temple, who conceived the motion of the planets without recourse to the notion of crystal spheres. He criticizes Maimonides for dismissing the sages' conception and for preferring the Greek philosophers: "Some think [correctly in our time] that the stars move in a circle without crystal spheres. In their opinion it was not good that the wise man of Israel set aside the original truth that was transmitted from the prophets to the sages of the Second Temple."[62] The new findings of Tycho showed that the ancient wisdom was correct, while in later generations this wisdom was lost. Later philosophers, Delmedigo stresses, erroneously held the premise that there is a physical, crystalline sphere, having forgotten the saying of the sages of the Second Temple.[63] Moreover, he shows his readers that the viewpoint supporting crystal spheres descended from the assumption that each sphere had a unique and particular motion that necessitated a separate soul. But this view, Delmedigo claims, was destroyed by Copernicus, who argued for unified motion that "religiously moves from west to east."[64]

Aristotelian metaphysics also troubled Delmedigo. His student Metz tells us that "this wisdom [Aristotelian metaphysics] lost its validity; many pens were broken and so much ink was spent on these debates. Some call those engaged with this wisdom metaphysicians, but my teacher [Delmedigo] was upset whenever he was named like that and said that he wished to be called a natural philosopher."[65]

Delmedigo stresses the invalidity of the metaphysics that based nature on four elements with primary and secondary qualities, and with eternal motion in space and time generated by the souls and intellects of the spheres. Instead, he suggests that "air is the thing that spreads and goes up and turns into ether. The planets move circularly against the elements, which by their nature tend to rest."[66] Once the planets move out of their place, they will all move linearly. Although he gets rid of Aristotelian natural circular motion and stresses instead a linear motion, Delmedigo still looks for an alternative cosmological explanation to replace the Aristotelian metaphysics. The spheres, he stresses, move uniformly with no particular souls as primary impulses, and metaphysics should be replaced with a more mechanical view of the universe in which mind and matter, or forces and bodies, make the machine work.

But this mechanical vision was not new. It came down to us by way of the prophets, who received this revelation of the secrets of nature from "Moses, whose intellect reached the highest, and some say that Thoth transmitted him the knowledge."[67] From the revelation of Moses and Thoth downward, the pure knowledge became corrupted and no longer pristine. Yet, in Delmedigo's view, the new cosmology showed that it merely had resurfaced. Accordingly, he draws a connection between the new cosmological discoveries and ancient theology and gives an account of the history of science as the revival of ancient wisdom: "In the time of our fathers there were many miracles, and the earth was filled with wisdom. Recently in the lands of Germany, Denmark, and England restless scholars have destroyed the barriers to knowledge and awakened ancient wisdom." For Delmedigo, the findings in cosmology and the discoveries in mathematics exposed godly origins, "like the table of the logarithms that was rediscovered."[68] He applied a cabalistic interpretation to the recent findings of Napier and Briggs in mathematics, viewing logarithms as numbers that refer, represent, and return to pristine numbers given by God.

Delmedigo entangled the hermetic critique—that Jews had lost primacy in natural philosophy—with his travels in the East. The new developments in Europe, made without Jewish contributions, highlighted the gap between a swiftly progressing Christian intellectual culture and a stagnant Jewish intellectual life. In turning to the Eastern Mediterranean for collecting Jewish manuscripts, Delmedigo struggled to rehabilitate the Jewish role in natural philosophy. In addition, because the new developments were not taking place in the Muslim world, the gap between Jews and Gentiles in natural philosophy was not yet evident there and carried no stigma. The decline in Jewish intellectual life implied that contemporary European Jews had lost track of an older, vibrant Jewish intellectualism. Reference to the Karaites and to the Jews of Cairo and Constantinople instigated a plea to take up ancient Jewish-hermetic revelation: the Jews had known it first.

Delmedigo was aware that his hermetic scheme clashed with the hegemonic intellectual discourse of rabbinical Judaism. Because he used rabbinical printing presses, he could not explicitly discuss "skeptical" works such as the natural philosophy of the Karaites. Worried about the charge of heresy, he used a method of self-censorship and only implicitly mentioned such esoteric sources. In several places in *Elim*, Delmedigo indicates that he was persecuted by rabbinical authorities and claims that he was even excommunicated. In the introduction, he muses about the biblical Joseph (who was persecuted by his brothers) and presents himself as the current Joseph that would save the Jews from further humiliation in contests of natural philosophy with Gentiles.[69] In another place, where he discusses the art of mechanics, Delmedigo mentions in a bitter tone that "if my sayings would enter the ears of observant and sincere men they would notice that my intention is good and my soul is clean from any impediment. And

I wish to have angels above me, running ahead of me against those who confront me. Then I will not be shamed by those hypocritical, envious, and crazy men. I hold my truth and I will not let it go because of those who criticize me, just as the brothers of Joseph were envious of him."[70]

Obviously, he toiled within certain cultural constrictions that had caused intellectual engagement with natural philosophy and the Karaites to be seen as a threat to mainstream orthodoxy. However, Delmedigo was not alone. He appropriated a skeptical and hermetic scheme that prevailed in the intellectual discourses in his home base—Venice.

A Venetian Nexus of Millenarian Christians and Skeptical Jews

The attraction to the Karaites was not only an internal Jewish matter, but part of a wider context. Specific external things—individual minds, books, and discussions—may have compelled Delmedigo to leave his home and family and travel to Karaite communities. At this time, discussions on the authority of tradition generated a wide interest among Karaite skeptics and millenarians. Christian millenarians stressed that the conversion of Jews to Christianity would happen only when Jews ended their diaspora and regrouped.[71] In the grand scheme, Christian theologians attributed a role to the Karaites. They were expected to lead the Lost Tribes to the Holy Land and to help restore universal religion and politics.

The renowned sixteenth-century Hebraist Guillaume Postel (1495–1581), one of the first Christians to be interested in the Karaites, spent three years in their community in Constantinople and obtained some of their writings. He traveled primarily not for scholarly but for political purposes. Linguistically skilled in Arabic and Hebrew, in 1536 he accompanied a delegation from Francis I, king of France, to negotiate an alliance with the Turks against the Holy Roman Emperor, Charles V (figure 18). Francis also gave Postel money to buy books in ancient languages. Venice was Postel's port of departure and return. Soon after his arrival there, before his travels, he worked within the general ambit of Daniel Bomberg's famous Hebrew press, which in 1528–29 printed the first Karaite work, an edition of the liturgy.[72]

On his return from Constantinople, Postel enrolled at Padua University.[73] His interest in the Karaites was part of an attempt to harmonize Christian, Jewish, and Muslim religions by integrating various sources of knowledge common to all three. In 1552 he wrote about the Karaites as a Jewish sect that broke off from the Samaritans, who were considered to be descendants of the Ten Tribes, and described their customs, linguistic mastery, and exclusive reliance on Scripture.[74]

Figure 18. Guillaume Postel in attendance at a reception for the French ambassador at the court of Süleyman the Magnificent. The French delegates were greeted ceremonially, and a military parade was arranged in their honor. There was a formal reception, during which Postel was allowed to kiss the Sultan's hand. The two sides exchanged gifts, and Postel presented a letter from Francis I, reaffirming the Franco-Ottoman alliance and urging Süleyman to attack the Habsburgs in Hungary. From Hazine, MS 1517. Courtesy of Topkapi Palace Museum Library.

Postel's sensibilities thus resonated with the hermetic tone prevailing in so many European centers of learning: its message was that once people of all faiths were brought into one religion, creation itself would be restored to its pristine glory in a grand *restitutio omnium*.[75] The "restitution of all things" played as a necessary prelude to the establishment of a universal monarchy. It could be achieved through careful use of old texts, either from the East or of cabalist and Karaite origins. Indeed, Postel's main interest in the East was the ancient Holy Scripture, but he did not deny the effectiveness of the late medieval Arabic astronomical texts that he brought back to Padua, texts that, apparently, Copernicus had used.[76] Postel paved the way for a larger European interest in the Karaites by other Christians and Jews, especially those concentrated in Venice.

Protestant theologians regarded the Karaites, because of their exclusive faith in Scripture and their rational critique and rejection of Jewish tradition, as the "true Jews" or "Protestant Jews." In the grand messianic Protestant scheme, the Karaites were expected to lead a general Jewish conversion to Protestantism and, by so doing, to accelerate the messianic era.[77] In the same vein, the English theologian John Dury corresponded with his friend Menasseh Ben-Israel, the printer of Delmedigo's works, regarding the possibility that the Lost Tribes were found in the New World.[78] In 1650, Ben-Israel published, in Latin, an apocalyptic work, *The Hope of Israel*, in which he stressed that Western and American Jews would gather in Egypt, and Middle Eastern and Asiatic Jews in Assyria.[79] Dury added that the Karaites would come from Russia, Constantinople, and Cairo and would lead the Ten Tribes to the Holy Land.[80] Moreover, the source for Dury's knowledge about the Karaites came from the cabalist Johann Stephanus Rittangel (1606–52),[81] who had studied in Constantinople, just like Delmedigo, spending twelve years with rabbis and Karaites.[82]

Furthermore, subversion of the authority of the Vulgate and growing interest in the "original" scriptures increased scholarly curiosity about Karaite biblical commentary. The Karaites' disregard for oral tradition and supposed reliance solely on direct, rational reading of Scripture attracted European Hebraists, who labored to reconcile Scripture with the deep changes in cosmology. The available Hebrew exegeses had already been studied and collated by Hebraists such as Postel, who subsequently turned to other ancient exegeses as possible sources of the Jewish commentators.

Protestant interest in the Karaites intersected with contemporary Jewish thinkers at various points, and directly and indirectly with Delmedigo himself. Hermetic Christians, men such as Postel and Rittangel, just like Delmedigo, spent some time in the East, especially among the Karaite communities of Constantinople. Their visits were also an opportunity for collecting Karaite manuscripts, as well as works of Muslim astronomers. The importation of these manuscripts helped generate a Hebraist scholarship that, in later generations, increased the Christian awareness of Karaite ideas. Moreover, Christian Hebraists who visited Constantinople also tended to sojourn in Venice, the gateway to the East.

Venice was Delmedigo's center of activity in 1613–14, when he was on his way from Padua to Crete. This visit is crucial for understanding what stimulated his interest in the Karaites and what propelled this Copernican to take a decision two years later to proceed to the East. It helps place Venice squarely within the context of Delmedigo, his hermetic passions, and the critique of rabbinical culture. The discourse about the Karaites in Europe and its connection to contemporary Jewish questions reached Delmedigo through the colleagues and masters who also wrote the approbations for *Elim*, and their credibility helped defuse any suspicion of heterodoxy. As it turns out, most of the writers of approbations for Delmedigo came from Venice—such scholars as da Modena, Luzzato, and HaLevi.

In Praise of Skepticism

Let us start with Leon da Modena, who seems to be a strong link between larger interests in the Karaites and Delmedigo. The appropriation of the Karaites by Christians, especially by Protestant messianic thinkers, caused Jewish rabbis to suspect certain Karaites of fostering heterodoxy. Joseph Kaplan has shown that many *conversos*, on their return to Judaism, carried new ideas about the Jewish oral tradition, thus prompting defensive actions to protect the rabbinical-oriented traditions.[83] Uriel da Costa had a direct bearing on da Modena and was one of the heterodox predecessors of Spinoza in Amsterdam. Da Costa had a positive view of the Karaite trend that went against Jewish oral tradition. Although he never officially aligned himself with the Karaites, da Costa found in their controversies with the rabbis support for his main argument that the oral law was merely human artifact, and he accused the rabbis of a "great heresy" for elevating oral law to the same level as the revelation of Moses.[84]

Da Costa's skepticism about the oral law reached the attention of Delmedigo's master, the rabbi Leon da Modena of Venice. In 1615, da Modena responded in a tolerant tone to a question from an Amsterdam rabbi who was rather anti-Karaite.[85] Da Modena's interests in Karaites, skepticism, and the debate about the Ten Tribes came through various sources and circumstances. In an undated letter to a friend, he writes that "he received a letter from Damascus about the Lost Tribes, [an issue] that has a renewing interest and approval these days."[86]

In another letter he mentions that he wrote a polemical book (no longer extant) against the Karaites.[87] He cites Karaite biblical commentary and praises himself for being "exceptionally fluent in the literature of the Karaites."[88] In the years pertinent to our discussion, 1614–15—when Delmedigo was in Venice and da Costa was spreading his brand of skepticism—da Modena, through fear of persecution, only seemingly confronted the Karaites.

In 1614–15, da Modena composed a work in Italian, *Historia dei riti hebraici* (History of the Rites of Present Jews), dedicated to James I of England.[89] The work was published in Paris in 1637, then printed in Italian in 1638 and in English in 1650. To the French translation, da Modena added two supplements that describe two Jewish sects—Samaritans and Karaites—and the basic ways in which they differed from mainstream Judaism. In the chapter on the Karaites, he reacts to the growing Christian interest in the sect and at one point mentions that Christian commentaries on the Karaites referred to them as the true representatives of the ancient Jewish religion.[90] What motivated da Modena to have a printing made in Paris was his acquaintance with Parisian Hebraists who were studying Samaritan and Karaite texts.[91]

Jacques Gaffarel (1601–81) was one of these acquaintances, a French Hebraist who was a friend of Tommaso Campanella in Richelieu's court. Gaffarel and da Modena had met in Venice and shared interests in Hebrew Scripture, hermeticism, and astronomy. In one of his own works, *Unheard-of Curiosities*, Gaffarel discussed creating horoscopes of the patriarchs; he argued that the Hebrew letters could be seen in the constellations and hence that divine knowledge was originally given to the Hebrews on Mount Sinai.[92] Gaffarel urged da Modena to print the work,[93] in response to publication of a work by the French prelate and theologian Jean Morin (1591–1659) concerning the Karaite commentary of Ahron Ben-Joseph (Ben-Josef).[94] Morin was highly active in converting Jews as part of the Christian messianic agenda, and he needed to show that rabbinical Judaism was severed from ancient Jewish traditions and that the rabbinical texts of the Bible had undergone serious change. The return to these traditions, he thought, would show the close relationship of Judaism and Christianity. Accordingly, Morin consulted manuscripts of the Samaritans that were brought to Paris by the French ambassador to Constantinople, Harley de Cèsy, and his friend Pietro della Valle; he also used the Karaite works brought by Postel.[95]

Gaffarel urged da Modena to go into print with *Historia dei riti hebraici*. The work presents Jews in general as acceptable to Gentiles, as nonsuperstitious, benevolent, moderate, and modest—probably to ease the criticisms being aimed at rabbinical Judaism by Christians, Karaites, and Jewish skeptics.[96] Moreover, da Modena dedicated a section to the Karaites and elucidated the essential differences between them and mainstream Jews, with no real signs of any bitter contemporary exchanges. Da Modena promoted a synthetic approach that called on Jews to engage with the social and intellectual world of the Gentiles and to revitalize their intellectual sources. He socialized outside his home without wearing a head covering, and he was engaged with many Christian scholars who came to study with him or to exchange views.[97] He collaborated with other scholars of alchemy and hosted them in his private alchemical laboratory in the basement of his family home.[98] On one of these occasions, his son Mordechai was exposed to toxic fumes and fell sick and died.[99]

Finally, da Modena deeply engaged with the sort of skepticism that was associated with Karaite thinking. He claimed to have obtained a provocative, anonymous manuscript entitled *Voice of the Fool* (*Kol sakhal*), which contained a harsh attack on rabbinical culture.[100] However, many of his contemporaries made the claim (supported by some modern scholars) that da Modena forged the name of the "unknown writer" and the dates, and that he himself was behind the opinions that he could not openly advocate. But more pertinent to our story, the work utilized a variety of Karaite skeptical arguments against rabbinical Judaism. In assessing the "mysterious work," da Modena says that the "enigmatic writer" heavily relies on Karaite writings.[101] The mysterious work argued that since the destruction of the Second Temple, the revelation of Moses had attenuated among the Jews, causing their political and intellectual decline. But the Karaites had followed a different course. Although the "enigmatic writer'" does not support Karaite ideas, he says that the Karaites still held to the correct revelation of Moses and did not fall into decline. On the contrary, they strengthened with time.[102]

The "enigmatic writer" also has something to say about the study of astronomy. Just like Delmedigo, he writes (and I suspect Delmedigo borrowed these very words) that at certain times, Jews had rituals involving dancing before the moon instead of raising their heads to the heavens and understanding the order of the stars and planets. And in a different place we find again the same words Delmedigo used, when the "enigmatic writer" states at the end of the essay that rabbinical Judaism "darkened the clear light of God's revelation, and I could not hold this evil problem in my stomach . . . I know that the rabbis and the heads of the academies would curse and excommunicate me and my work. However, maybe, one single person in a generation, who is clean from this foolishness, would find my arguments appealing."[103]

We find another connection in da Modena's approbation for *Elim*. Da Modena mentions Delmedigo's trip from Padua to visit him in Venice. But more importantly, he writes: "the righteous among our people, who would shut their eyes from natural philosophy and metaphysics, only cover their stupidity and foolishness with pretentious holiness. I wrote against them in my work *Sha'agat arieh* [the published version of *Voice of the Fool*]."[104] Da Modena is implicated as the "enigmatic" authorial voice of the heretical work, indicating the convoluted art of writing that a skeptical Jew had to use in discussing Karaite works.

Da Modena, Delmedigo, and other critical, if not skeptical, rabbis were not the only ones to stress the decline of Jewish intellectual culture and to preach in favor of cultural, intellectual, and economic assimilation with Gentiles. Another approbation writer, a senior member of the Jewish community in Venice and a friend of Delmedigo, was Simone Luzzato (d. 1663). In *Elim*, as noted earlier in the chapter, Delmedigo often stresses that the only intellectual Jews he met who could understand and cultivate the new findings in astronomy and mathematics were the Karaite Jacob Iskandrani of Cairo, the Karaite Zerah Ben-Natan of Troky, and the rabbi Simone Luzzato of Venice.[105] In a different work, *Spring of Life* (*Ma'ayyan ganim*), Delmedigo states that no one in the Jewish world really understood the new astronomy "apart from one notable man from Venice, whose fame has reached to the top level among Gentile scholars, rabbi Simone Luzzato, knowledgeable in all the parts and details of knowledge."[106] Luzzato was an exceptional case for Delmedigo, because he was well educated in "external" (that is, non-exegetical) studies, including politics, economy, and natural philosophy. Moreover, he also stressed the cultivation of contemporary natural philosophy as an urgent necessity for Jews.

Luzzato promoted the belief that Jewish writers should engage intellectually with Gentiles. He composed a utopian essay, *Socrate overo dell'humano sapere* (1651), describing a scientific society on the island of Delphi, in which intellectual freedom would help men acknowledge the ultimate truth, accessible to them by a combination of reason and divine inspiration. The society would be solely dedicated to proving revelation through the application of reason to the exploration of nature.[107] However, Luzzato is mostly known for his work *Discorso circa il stato degli Hebrei* (1638), in which he promotes the incorporation of Jews into Gentile society.[108] He argues that the universal character of the religion of Moses might benefit both Jews and Gentiles in any society where they found themselves. Incorporation would bring to Jews stability and rights without persecution, and they could work on a revival of their wisdom.

Like Delmedigo, Luzzato believed that before the Greeks, Jews had been masters of natural philosophy. However, in contrast to Delmedigo, he thought that the decline of Jewish intellectual life occurred because of exile and political insecurity: "virtue and scholarship exist only with a comfortable life."[109] Accordingly, ever since the destruction of the Second Temple, Jews had been in danger of "declining into total ignorance and only the reading of the Bible made them modestly use their mind and curiosity in the science of nature."[110] Luzzato mentioned three great Jews who combined reasoned revelation with the secrets of nature: "Moses, the head of the prophets, Philo Judaeus a genius philosopher before the decline of the Jews; and Maimonides the genius philosopher after the decline of the Jews," all associated with Egypt.[111] The fear of Jewish intellectual decline impelled Luzzato to encourage study of the new astronomy. In praise of Delmedigo's *Elim*, Luzzato writes: "my heart was joyful to discover these wisdoms [in Delmedigo's book], it would close the lips of those Gentiles who arrogantly state we have no wisdom. These days the scholars of Greece and Rome would say: they have just the same hearts as we do."[112]

According to Luzzato, works in contemporary astronomy, like those of Delmedigo, were important not only to restore Jewish intellectual pride, but actually as the highest level of worship of God, because "it is a commandment from the Torah to contemplate nature in order to come to the closest truth about the creator. Therefore it is an imperative to study astronomy, for it will help us determine Time, and also [will be] a gate to and an introduction to the omnipotence and omniscience of God."[113]

Luzzato devoted only a small portion of his book to the question of the Karaites. To him, the rejection of traditional rabbinical readings caused unresolvable arguments and made the Karaites an ungrounded sect. However, he praised them for their metaphysical arguments that the soul is imperishable and, as in hermeticism, that spiritual angels transmitted the word of God to Moses.[114]

Finally, we consider the approbation for *Elim* written by Jacob HaLevi. He was the son-in-law of da Modena and the student of David Ben-Shushan, the mathematician and colleague of the Ottoman astronomer Taqī al-Dīn. HaLevi repeats the arguments of Luzzato and da Modena that "this work will comfort us from the upsetting things that were said by the Gentiles about the righteous people [the Jews], [implying that] we have lost the wisdom and reason of our ancestors."[115]

Venetian scholars of that time were engaged with the problem of skepticism and Karaite arguments. However, the defenders of rabbinical authority, such as da Modena and Luzzato, still generated strong criticism against the stagnation and isolation of Jewish life. They were sources for Delmedigo's exhortations to Jews and anchors for his unique criticism of the Jewish role in natural philosophy. On the one hand, Delmedigo adopted the skeptical and Karaite notion that the revelation of Moses was ultimately superior to any later teaching of the rabbis. On the other hand, he criticized the narrow-mindedness of rabbinical intellectual culture, which impeded the development of natural philosophy in the academies.

Later radical Jewish critics talked about the fall of the Jews after the destruction of the Temple and the need to revitalize Jewish social and cultural life. The revitalization would require two complimentary stages. First, given that the source of civilization was the revelation of Moses, the Jews must recover this pristine knowledge and thus reform their lives and integrate into society and into intellectual and cultural currents. Second, Gentiles must open doors for Jews, who could serve as a link to ancient sources of natural philosophy.

WE HAVE TRACED DELMEDIGO'S WORK through the clues provided in his writings and from evidence provided by his position in a nexus of scholars. While studying in Padua, he was convinced by the Copernican cosmology and, at the same time, grew increasingly disturbed that, for generations, Jews had not made significant contributions in natural philosophy. During his visits to Venice, he was exposed to skeptical and Karaite arguments against rabbinical intellectual culture, and he appropriated these arguments, with a hermetic-cabalist twist, into his own views on Jewish intellectual history. The poor state of natural philosophy among the Jews, he concluded, was a result of two things: the rabbinical methodology that elevated oral law above Mosaic revelation and the priorities of printers. The Hebrew presses preferred religious works and by so doing excluded many Hebrew manuscripts of natural philosophy that in earlier times, Delmedigo supposed, had circulated within Jewish intellectual culture. His solution for promoting Copernican cosmology among Jews and for incorporating Jews into the contemporary culture of natural philosophy was embodied in the utopian Elim—the very origin of revelation. He would show that the new cosmology exposed shards of the unitary revelation of natural philosophy given to Moses. Consequently, he collected manuscripts of the Karaites and of Ibn-ʿEzra in the East, where he hoped to find the hermeneutic threads that led back to the *corpus hebreicum* and to Elim.

Delmedigo's esoteric writings offer a possible direction for further research. Until now, intellectual historians have looked for coherent structure in his thought, but have overlooked the mechanisms of the print culture and the ways in which only certain works made it into print and were or were not accepted as authorities. Seeking coherence of thought and taking texts as self-contained and fixed can be a deceptive approach. In an important comment made by the printer Menasseh Ben-Israel at the end of *Elim*, we find additional information on the process of transforming manuscripts into books in early modern times. Ben-Israel had urged Delmedigo to publish his work because of a lack of knowledge, in Hebrew, on recent natural philosophy. Delmedigo was rushed and had neither the opportunity "to let his thought marinate in his mind" nor the time to revise the first draft. Ben-Israel thus felt compelled to write that "many of his [Delmedigo's] tales are obscure, and there are also many mistakes in the text. Whoever does not understand the text or would find a mistake, after my proof editing, [should excuse me] since the text came to me with inaccessible handwriting and as drafts full of mistakes. Therefore, I beg before all the scholars not to blame me and not to lay the sin on me."[116]

Delmedigo's lack of clarity and miscues with his printer produced an almost inaccessible work that only few in Amsterdam could grasp. One of the few who did delve into Delmedigo's esoterica was Benedict Spinoza, a reader of *Elim*. According to J. D. Ancona, Delmedigo was a source for many of Spinoza's arguments about the decline of the Jewish people.[117] One might suggest that the Venetian circle that called for a Jewish radical enlightenment was the precursor of the Radical Enlightenment promoted by Spinoza.

Converting Measurements and Invoking the "Linguistic Leviathan"

JUST BEFORE HIS DEATH in 1652, John Greaves (b. 1602), a professor of astronomy, published *Astronomica quaedam ex traditione Shah Cholgii Persae*, a bilingual, Persian-Latin edition of a late fifteenth-century astronomical work from Persia. To this work he attached a Persian-Latin astronomical dictionary, which was addressed to Latin-readers. The inspiration for this linguistic astronomical project came in 1637, when Greaves left his position at London's Gresham College to travel for a few years in the Near East. In his travels, he conducted astronomical observations with local astronomers, measured the pyramids, looked for the perfect ancient cubit, and collected Arabic and Persian manuscripts on astronomy and geography. On his return to England in 1640, he was appointed to the Savilian Chair of Astronomy at Oxford University and, while there, published works on the pyramids, Persian language, ancient measurements, and Arabic geography—all unexpectedly addressed to Latin-readers.

By the time *Astronomica quaedam* was published, Greaves was at the end of his academic career and near the end of his life. He had been ejected from Oxford, along with all other royalists, during England's Civil War, and his two major patrons, Archbishop William Laud (1573–1645) and Archbishop James Ussher (1581–1656), were either dead or unable to protect him from the parliamentarians who had taken over the government. Not only did the parliamentarians and their sympathizers reject royalist claims, but they also espoused a new experimentalist view of nature that advocated firsthand experience while disregarding traditional authority and history. Stranded and miserable, Greaves spent the rest of his life preparing publications on astronomy and geography that searched for true cosmology in medieval Islamic sources. At a time when traditional Islamic astronomy was pushed aside, Greaves—perhaps surprisingly—had in mind readers who would be interested in a Persian-Latin dictionary of an astronomy that had lost its historical authority.

The interest in apparently anachronistic astronomical sources may seem extraneous to the development of post-Copernican astronomy. However, the linguistic reform in astronomy was situated at the heart of various cultural and political trends. The political events that took place in mid-seventeenth-century England sparked controversies over philosophy and political theology that captivated scholars such as Greaves, who started debating the question of philosophical demonstrations and authorities of truth in natural philosophy. The work of Steven Shapin and Simon Schaffer throws light on how politics was reflected in the approaches of Robert Boyle and Thomas Hobbes toward epistemological problems in natural philosophy.[1] The question at stake was how to promote and integrate a community of natural philosophers that could resolve controversies in the best possible, most credible, way. On the one hand, we see Boyle, the parliamentary sympathizer, placing experiments under the scrutiny of trustworthy witnesses; on the other, we have Hobbes, the royalist, who held that there is a need for a "Leviathan," a "meta-philosopher," that might resolve controversies and establish credibility in natural philosophy.

Greaves also played a central role in the establishment of English Orientalism. Historians working in that field tend to believe that the search for answers in the Near East was part of an increasingly global interest in the "Orient."[2] Here, however, we probe the seemingly anachronistic interests in Arabic and Persian sources of astronomy. Why, after the new Copernican and post-Copernican astronomical systems, which invalidated ancient Greek and Arabic astronomy, would English astronomers find Arabic and Persian sources to be so crucial to their work?

The answers will emerge from a deeper social story about John Greaves (1602–52). The appeal of the "Near East" was not primarily to learn about the "Near East," but was part of a search for the "primordial sources" needed to resolve various contemporary controversies in both political theology and natural philosophy. In his travels, Greaves interacted with material objects such as ancient monuments, coins, and manuscripts. As a by-product, he informed local astronomers about the new cosmologies and taught them the observational techniques of Tycho Brahe. Greaves's travels—observing eclipses, measuring latitudes and ancient monuments, collecting and comparing texts—were set against the backdrop of a historical a priori: a "Linguistic Leviathan" (to coin a phrase). He searched for a primordial, divine language of words and units of measure that could be applied to descriptions of nature. The quest was motivated by his belief that multiple cosmologies arose because of the deterioration of astronomical terminology: in the process of translation, the true meanings of words were lost.

His methodology aimed to purify language by returning to the one, primordial language to resolve controversies. In a sense, the quest for a Linguistic Leviathan that occupied Greaves and his colleagues reflected a search for a "true science" that could simultaneously work as a gateway to religious truths.

To fully understand the cultural lineages of Greaves's methodological principles, we can examine texts of three different types and formats. The first type elaborates his interest in measurement, nomenclature, and antiquity—his scientific works on Islamic sources of astronomy and on ancient measurements. The second type consists of personal correspondences, including letters and essays composed for his patrons—archbishops Ussher and Laud. These documents help us understand the direct instructions and the subtle motivations for his voyages and writings. They were mainly gathered by Thomas Birch and published in 1737 as the *Miscellaneous Works of Mr. John Greaves*,[3] and a few are also found in the collected works of Ussher and Laud. The third type of source is the writings of Greaves's colleagues and foes in the academic world of Oxford that situate the Linguistic Leviathan within the contemporary controversies in political theology and their effect on adversarial camps in natural philosophy.

Ancient Meanings and the Purification of Astronomical Language

In chapter two, we saw that Pietro della Valle, like John Greaves, took up the difficult task of composing bilingual essays on astronomy. Yet the two men's reasons for doing so were strikingly different. Whereas della Valle translated the work of Christopher Borrus from Latin to Persian to introduce the Tychonic world system to, as well as attempting to convert, Mullah Zayyn al-Dīn al-Lārī, Greaves wanted astronomical terms in Arabic and Persian to become known to his European peers. Consequently, we face the question of what use post-Copernican European astronomers might have had for such a lexicon. The introduction to *Astronomica quaedam* supplies an explanation.

> A little less than four hundred years ago, Gerard of Cremona [the famous translator of the twelfth century], a man who was exceptionally skilled in the Arabic language though not as well versed in astronomy, published [translated from Arabic to Latin] theories of the planets. Regiomontanus . . . was the first to rebut his errors, which had been accepted here and there in the schools and had been rashly approved by professors who were unskilled in the arts. In fact, even a little before his [Regiomontanus's] time, Georg Peurbach, an eminent astronomer and Regiomontanus's master . . . wrote his book *Theoricae novae*

planetarum, by which he facilitated the reading of Ptolemy and the ancient astronomers.[4]

Gerard of Cremona was the first to translate many philosophical works from Arabic to Latin, including Ptolemy's *Almagest*,[5] but his lack of proper understanding in astronomy and astronomical nomenclature resulted in the production and canonization of astronomical texts that were an imperfect Latin representation of the originals. Peurbach's and Regiomontanus's criticism of linguistic errors in the Latin translation of the *Almagest* inspired an investigation of the original Greek works and was a necessary and welcome step forward. However, Peurbach did not have access either to Ptolemy's *De hypothesibus planetarum* or to Arabic works such as those of al-Battānī and al-Ṭūsī, which introduced Ptolemy's short work and developed a nomenclature of astronomy. Greaves gives credit to Peurbach for at least having written a summary of the particularly elusive aspects of the science.

> Therefore Peurbach should be praised for being the first after the rebirth of learning in Europe who wrote a short introduction to the most perplexing parts of astronomy. Since the time of Peurbach there have been published several treatises upon the elements of astronomy, or commentaries upon that writer [Peurbach]. Amongst these the most eminent are Erasmus Reinhold and Michael Maestlin, the latter of whom is frequently recommended by Tycho Brahe. But even these writers have not explained everything to the extent that an attentive reader could wish. For, we find in them a fair number of barbarous terms unpleasing to the Latin ears, but everywhere used in the writings of astronomers, the origin of which [terms] is a matter of interest for the Republic of Letters.[6]

But why was it necessary to understand Muslim astronomical sources in their original languages? Greaves understood that the Latin astronomical nomenclature used in Europe was formed by incorporating Arabic terms. From the Alfonsine tables, which supplied a mass of exotic words in the Latin texts on astronomy, "came the words Juzahar, Zenith, Nadir, Buth, with an unusual variety of others, either taken from the Arabians, or formed in imitation of them." He gives further examples, stating that even the Arabic terms were probably corrupted meanings of either Greek or Mesopotamian words: "for it happens in science as in names of countries and places, that what has once been commonly received, will be obscured by the length of time."[7]

Underlying the corruption afflicting astronomical terminology lay a decay in the true meanings of words, which had once resided perfectly within a "universal language." In tracing exotic words to their primordial sources, Greaves chose astronomical terms from the commentaries of Mahmūd Shāh Khaljī, hoping that the reader would be doubly benefited.

> For those who are familiar with astronomy will see the origin of several words without which the tables used, at one time by the Arabians as [at] another by the Persians and Indians, cannot be understood; and [will] perceive that the celestial hypotheses of those nations are exactly compatible to those of Ptolemy [see figure 19]. We have them [the words] briefly and clearly explained here, and adapted to the motions of the planets from the accurate observations of al-Ṭūsī in the city of Maragha. Those, likewise, who study the oriental tongues, will be pleased to see a book published in the genuine Persian language.[8]

For various reasons, Greaves preferred to present the little-known fifteenth-century Shāh Khaljī's astronomical tables rather than simply make a direct translation of those of al-Ṭūsī. We know that Mahmūd Shāh Khaljī was the ruler of Malwa Persia during the years 1435–69. His work *Zīj al-jām'i* (The Universal Astronomical Tables) canonized the astronomical tables of the two great Islamic observatories: Marāgha in the thirteenth century and Samarqand in the fifteenth.[9] Fifteenth- and sixteenth-century works of Islamic astronomy were, in fact, innovative in their observations and in adding mathematical solutions to the problems of planetary motion, such as Khafri's four explanations for the Ptolemaic model of Mercury.[10] Greaves chose later work, such as that of Shāh Khaljī, because it standardized and canonized terms and gave general, conclusive summaries of medieval astronomical tables. His interest in such texts shows that Europeans used not only prominent ninth- to thirteenth-century Islamic sources, but also later Muslim astronomical achievements. Even after Copernicus, astronomers who invoked antiquity to confirm the Tychonic system were dipping into contemporary Islamic texts, looking for a projection of the past onto modern theories. Thus, instead of compiling a summary of many primary sources of Muslim astronomy, Greaves benefited by using a work (Shāh Khaljī's tables) that carefully smoothed out the essential points of Muslim astronomy. It enabled him to avoid the mistakes made by copiers over the years and to present, more efficiently, a precise account of standard Muslim astronomical terminology. Islamic commentaries could facilitate solutions to contemporary astronomical problems. Dissatisfied with misused astronomical terms, Greaves called for a return to the old Arabic and Persian astronomical literature through which one

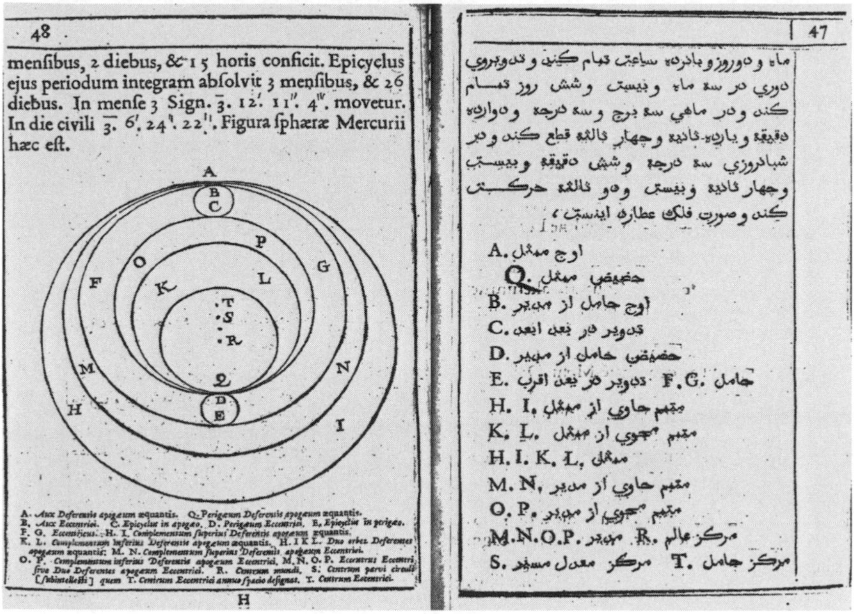

Figure 19. Sample pages from John Greaves's *Astronomica quædam*, clarifying technical astronomical nomenclature. From Greaves, *Astronomica quædam ex traditione Shah Cholgii Persae: una cum hypothesibus planetarum* (1652), QB41. Courtesy of The William Andrews Clark Memorial Library, University of California, Los Angeles.

could reclaim precision in language, in the numerical data of astronomical tables, and in units of weight and length.

Yet, Greaves's travels in the Near East were more than a search for texts. In addition to collecting manuscripts, he carried on professional conversations with local astronomers. While in Constantinople in the 1630s, he looked for approval among scholars for the Tychonic system, but not only by consulting primary sources (figure 20). He also conducted dialogues with local astronomers about the two chief systems of the universe, although he apparently did not appreciate their opinions: "Turkish astronomers of no means or skill" stressed their agreement with "the observations of Tycho Brahe, a name that in these regions and ours had gained an overall fame [*nam in eas regions e nostrabitus ejus unius fama pervenit*]."[11]

Greaves's preparation of an astronomical dictionary seemed to be part of a larger project. In the preface to a treatise on geography, *Binae tabulae geographicae*

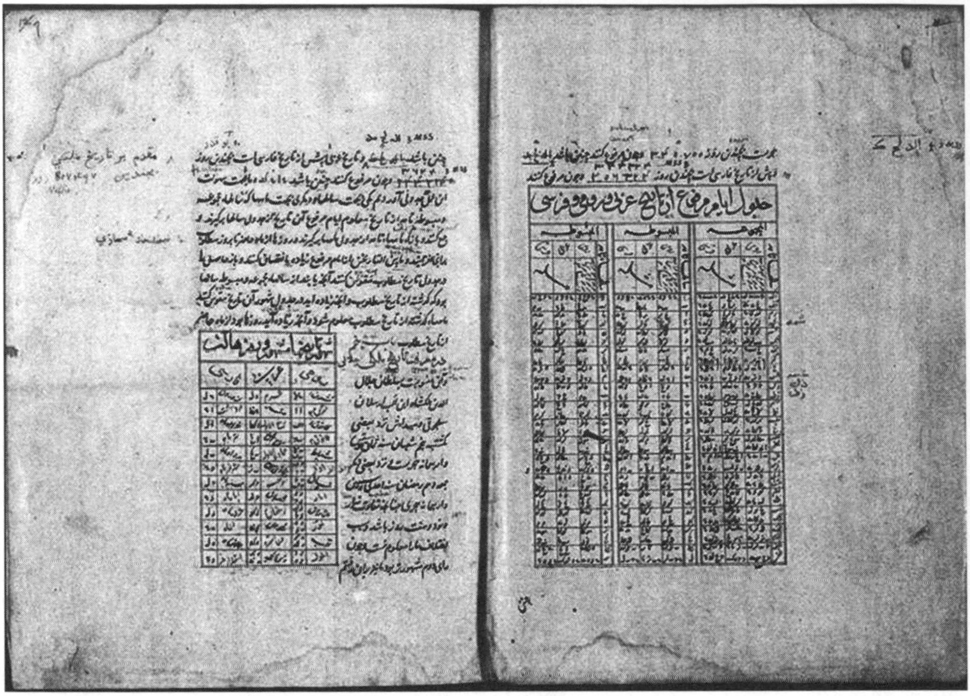

Figure 20. Text pages with calendric tables in one of the manuscripts Greaves found on his travels and brought back to Oxford. From *Zij al-Jadid al-Sultani Ulugh Beg*, MS Greaves 5, Bodleian Library, University of Oxford.

une Nassir Eddini Persae, altera Vlug Beigi Tatari, which he attached to his main work on astronomy, Greaves mentioned that some scholars implored him to undertake the editing of the "Geographical Canon" of Abū al-Fidā and to satisfy the European demands "for a usable edition of al-Fidā, by which, many argued, geography would be illustrated properly."[12]

Monstrous Copernican Astronomy

The need for precision of data and accuracy of text became urgent. Greaves and his social peers were concerned with certain astronomical-geographic problems, ones that inspired mathematical voyages. Imprecise astronomical tables, they argued, had resulted in "monstrous" (Copernican) astronomical theories. Greaves came to such a conclusion after a close and idiosyncratic reading of Copernicus's *De revolutionibus*. In a private collection in Oxfordshire, Owen Gingerich found a copy of *De revolutionibus* with Greaves's marginalia, which formed the great

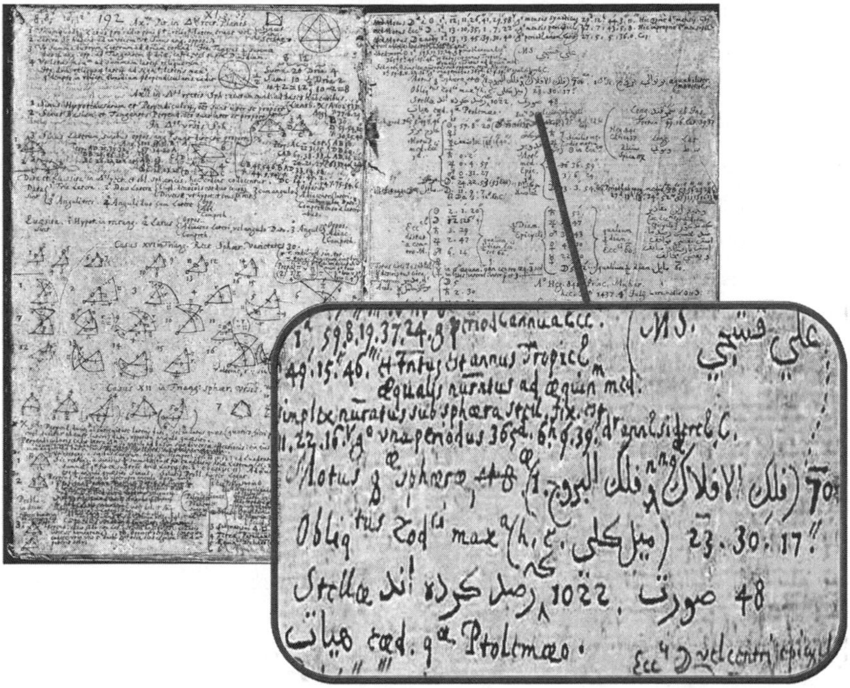

Figure 21. Greaves's marginalia on his copy of Copernicus's *De revolutionibus*, first edition (1543). In the enlarged section, Greaves compares data from the Ptolemaic tables with those of ʿAlī Qūshjī, a fifteenth-century astronomer and member of the Samarqand Observatory. This copy was in the library of the Earl of Macclesfield; it was removed from Shirburn Castle in 2004 and sold by Sotheby's to an anonymous buyer. © Sotheby's.

bulk of the annotations on this copy, with some mathematical and astronomical notes and planetary parameters in Arabic and Persian (figure 21).[13]

The gradual acceptance of the Copernican model, facilitated by Galileo and Kepler, put Tychonic astronomers such as Greaves in an apologetic position. He claims in his *Binae tabulae* that "if these Islamic sources [which he had been collecting in the Near East] had been known to the Europeans in the preceding ages, those monstrous hypotheses like Copernicus's theory of eight heavens [*portentosae hypotheses octavi coeli*], long before introduced by Thabit Ben-Qurra, would have been exploded."[14] He also refers to Copernicus's use of the theory of precession of the equinoxes. The earth's slight, constant wobble makes the axis rotate, thus changing the sited position of the celestial pole. Observations repeated over very long periods disclose the pattern of precessional motion; as the centuries pass, the celestial pole moves gradually through the stars in this circle, at a rate of 0.7 degrees per century, and completes one revolution every twenty-six thousand years. The consequences of correctly computing precessional motion were radical, affecting the frame of reference in which longitudinal motions occur. Moreover, the apparent position of the pole would be changing over time, which means that the timing of seasonal nodes, the North Star (polestar), the zodiac, and geographic coordinates were changing as well.

A royalist such as Greaves appealed to the divine political authority of the king and the unchangeable authority of ancient wisdom. He maintained a solidly geographic understanding of Copernican precession, "that the poles of the world changed their sites, and subsequently all countries [have changed] their latitudes,"[15] and thus time and place in antiquity were impermanent. An unwanted result would be that Holy Scripture could not remain eternally authoritative on any reference to the location and time of events. The radical implications of Copernican precession disturbed the political and natural order.

All of this leads us to the logistics and itinerary of Greaves's travels. If ancient latitudes for cities recorded by the Greeks and Arabs were the same as modern latitudes,[16] without even a few degrees of change since the time of Ptolemy, then, in Greaves's thinking, the poles of the earth do not change their locations. For this purpose, he visited the ancient cities of Alexandria, Rhodes, Constantinople, and Baghdad and calculated the local latitudes by measuring the altitudes of the celestial equator or the celestial North Pole. Apart from measuring latitudes, he also had an ambitious astronomical project concerning the longitudes of these ancient cities. He traveled to arrange simultaneous, comparative observations (in Alexandria, Baghdad, and Constantinople) of a lunar eclipse predicted for December 10, 1638.[17] By training local astronomers—mainly Europeans living in the Near East—and equipping them with modern instruments and techniques, he planned to measure the altitude and azimuth and, most importantly, to record the local time of the event, to determine the time differences between the cities and thus their longitudes.

Through his travels, Greaves hoped to put the numerical data of Ptolemy, Copernicus, and Tycho to the test of precision. That his observations of the eclipse and his other observations in Rhodes and Alexandria resulted in numbers closer to the Ptolemaic than to the Copernican tables was further proof of the imprecision and refutability of the Copernican system. Whether heliocentrism turned out to be true or false was not as important for Greaves as the deeper implication—that Copernicans believed a theory in which ancient knowledge had no value and the natural order was subject to changes in astral-chronological timing. The royalist professor of astronomy adhered to a deductive political order headed by the king. His way of saving the old order of nature was to criticize the Copernican system, not on theological, physical, or mathematical grounds, but at the level of language and numerical precision. Thus, Greaves had a unique and unconventional take on the Copernican system. By stressing that his measurements of latitude did not detect the expected difference (based on a polar shift of 0.7 degrees per century) from Ptolemy's accounts, he undermined Copernicus's use of precession and Copernicus's entire system.

Greaves's work focused primarily on criticism of radical developments in astronomy and especially of the Copernican system, which he found repellent. In his view, two new astronomical theories were established in less than a century simply because of the corruption of astronomical nomenclature and the loss of original texts from antiquity. The radical disagreements in astronomy were, in fact, a result of cumulative mistakes in translations, transcriptions, and usage of technical astronomical terms. The works of Ptolemy and the wisdom of the ancients had passed through too many mistranslations, a process in which, just as for language in the Tower of Babel, a universal astronomical nomenclature and precision and true knowledge were lost.

From Language to Numbers: Restoring Precision to Ancient Measurement

Greaves castigated the language of astronomy and demonstrated that translations had distorted ancient knowledge into monstrous theories. His metrological research represents a unique consideration of the ancient language of natural philosophy that resonated with royalist political theology. Knowing only his bilingual dictionary, one might think Greaves believed that remnants of decayed nomenclature could be found in the late medieval Arabic and Persian astronomical works. However, through his *Pyramidographia*, in which he recaptures the measurements of the pyramids, and through *A Discourse of the Roman Foot and Denarius*, on the most precise classical measurements of cubit and foot, he searched for a metrological datum from antiquity—long before Islam.

Greaves partly followed the argument of his patron Archbishop Ussher, which held that sources of Arabic and Persian could facilitate knowledge about the Bible and natural philosophy.[18] Astronomical writings in Arabic and Persian were important not in themselves, but as carriers of the traces of a universal numerical knowledge of the dimensions of the cosmos that existed at some point in antiquity. Greaves's religious notions concerning Muhammad were not favorable. He referred to Muhammad as a false prophet and condemned him for delivering lies. In the essay on the pyramids, he states that Arabs excelled in the speculative sciences but not in histories and events in ancient times, and hence their account of the sources and meanings of the pyramids should not be considered reliable.[19] As Greaves moved from Arabic-Persian texts on astronomy toward his work on ancient Egyptian, Hebrew, and Chaldean sources, he did not believe that the Persian and Arabic texts were necessarily accurate representations of the original works of Ptolemy and the ancients, merely that they were less corrupted than the translations into Latin. To rehabilitate universal knowledge required deep skill in the knowledge of ancient Egypt and Chaldea. Thus, he targeted not Islamic intellectual traditions per se, but the data contained in Muslim texts.[20]

In Egypt, Greaves searched for two types of source material: the physical monumental remains of the tombs and sculptures of Ptolemy and Hermes, and the secrets of the construction and measurements of the pyramids.[21] He was chiefly preoccupied with the question of why the ancient Egyptians built the pyramids (figure 22). Construction of the pyramids was a matter of the highest form of truth, because the ancient Egyptians invented astronomy and the zodiac signs. Therefore, "seeing [that] in these [pyramids] all things are made, and that the coming of the sun, which is as it were a point in respect of those signs, is the cause of the production of natural things, and its departure the cause of their corruption, it seems very fitly, that by a Pyramid, Nature, the parent of all things, may be expressed."[22]

There were various arguments about Egyptian knowledge. In one view, the Egyptians were excellent geometers and astronomers, and therefore traditional scholarship has believed that the pyramids expressed the first principles of Egyptian mathematics. Scholars mentioned, with a Pythagorean undertone, that as excellent arithmeticians, the Egyptians made the pyramids to represent the mysteries of pyramidal numbers; or that as philosophers in the natural science of optics, they made the pyramids to demonstrate the interplay of shadow and rays from luminous bodies. Although Greaves accepted such explanations, which seem to connect measurement and geometric structure with research into natural philosophy, he argued that they were inadequate. For him, the pyramids were symbols of unchanging historical authority. The Egyptians made the pyramids because "they apprehended it [the pyramid] to be the most permanent form of structure, as in truth it is, for, by reason of the contracting and lessening of it at the top, it is neither over-pressed with its own weight, nor is it so subject to the

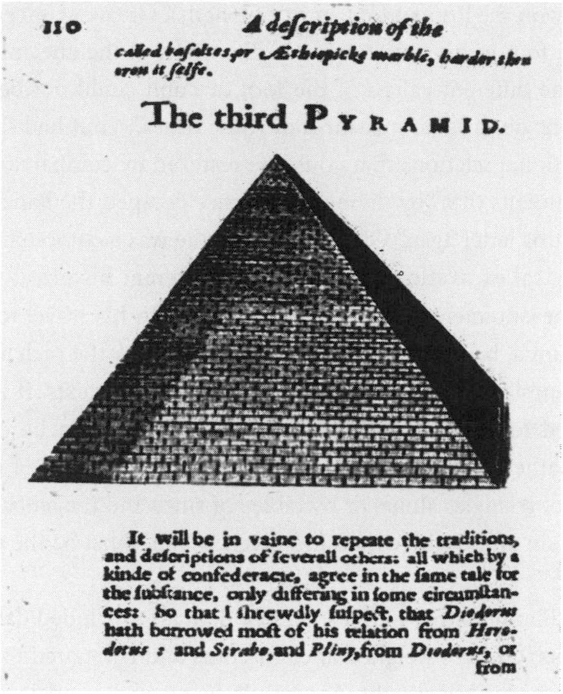
110 A description of the
called basaltes, or Æthiopick marble, harder then iron it selfe.

The third PYRAMID.

It will be in vaine to repeate the traditions, and descriptions of severall others: all which by a kinde of confederacie, agree in the same tale for the substance, only differing in some circumstances: so that I shrewdly suspect, that *Diodorus* hath borrowed most of his relation from *Herodotus*: and *Strabo*, and *Pliny*, from *Diodorus*, or from

Figure 22. Illustration from John Greaves's *Pyramidographia; Or, A Description of the Pyramids in Ægypt* (1646). Courtesy of The William Andrews Clark Memorial Library, University of California, Los Angeles.

sinking in of rain, as other buildings."[23] Thus, the structures were relatively permanent: they still existed and were beyond the natural process of degeneration and decay that affected other buildings and things in nature. The measurements of the pyramids, as they stand today, are the same as when they were built. This permanence, for Greaves, pointed to the possibility of recapturing precisely the Egyptian metrological system. He judged the pyramids to be embodiments of divine ancient knowledge expressed in geometric forms that conveyed meaning about the structure of the universe. The key lay in grasping primordial, ancient measurements.[24]

The measurements used by Egyptians in constructing the pyramids were not the only concerns about ancient measurements. In accord with the significance that Laud attributed to the Jewish Temple, which carried "the original copy of the law, the word of God written in tables of stone,"[25] Greaves saw in the measurements of the pyramids traces of the precise data for the sacred cubit, the unit of measure used in constructing the Temple.

In his essay on the Roman foot (unit of length), Greaves conceived the measurement of a foot in his own time as a distortion of the ancient "cubit of the sanctuary." The different values of the foot or cubit could not be reduced, "as mathematicians observe, to measures of these times,"[26] but had their own universal proportional relations that could be restored by comparison of different ancient monuments that "by divine providence escaped the hands of ruin and continued to this latter ages." Greaves's technique was a combination of textual and archeological excavations. He looked at different historical texts that described the measurements of extant monuments. In his travel to Italy, Egypt, and Mesopotamia, he measured several monuments by the inch and converted his results to cubits or feet, cross-checking with ancient texts. It was as impossible, he argued, to recover the Roman foot from naked, unanchored description "as it is for mathematicians to take either distance or altitude of places, by the propositions of triangles alone, or by tables of sines and tangents, without having some certain and positive measure given, which must be the foundation of their enquiry."[27]

The rehabilitation of the Linguistic Leviathan also included data concerning ancient weights. Greaves weighed ancient coins and compared the results with textual evidence of specie weight. As a result, he reconciled ancient weights with their contemporary counterparts. By obtaining the precise weight of the ancient Roman denarius, and by using archeological findings and textual testimonies in Hebrew, Arabic, Persian, Samaritan, and Greek, he established a precise weight of the Hebrew *shekel* (שקל), which was used in weighing the instruments of the Temple. The divine measures and weights were made "for high and sacred use, were kept in the Sanctuary, for God himself made this one . . . And it is no wonder that God, who so much hated a false balance, and a false measure, should commit the charge of these to the priests, as things most holy."[28] The size of the Temple and the weight of what it contained constituted something divine: "the Cubit of the sanctuary was taken from the cubit of Adam, he being created in an excellent state of perfection."[29]

The recovery of primordial nomenclature and units of measure also reconciled the ancient with the new "values of precision" and thus formed "a responsible, non-emotional and objective" language that would serve as a new, reconciling public language for natural philosophers and theologians. Standardization of units of measure also reflected a desire to centralize bureaucracy and political power and to regulate human resources.

Appeals to Authority for the Resolution of Church and Civil Problems

Arguments for recovering ancient units of measure prevailed in social and cultural surroundings. The late 1620s formed Greaves's intellectual worldview. After training in Greek and Latin classics, he studied natural philosophy and mathematics in a circle of colleagues working in mathematics and astronomy, including Henry Briggs, John Bainbridge, and Peter Turner. Within this circle he expressed discontent with the various opinions in astronomy, from Peurbach to Kepler, and went back to old sources and read closely in Greek, Arabic, and Persian sources.

His reputation grew. In February 1630 he was appointed professor of astronomy at Gresham College in London, and at the same time his friend Peter Turner introduced him to the Archbishop of Canterbury, William Laud, who at that time was also the chancellor of Oxford University. Greaves then decided to travel to the Near East. In 1635 he left for Leiden, Paris, and Florence, and he arrived in Constantinople in August 1638. Anthony Wood's collection of biographies of Oxfordian figures, published in 1689, informs us that Greaves's grand design was to visit the Near East and that the patronage of Archbishop Laud made it possible: "his Grace sent him [Greaves] to travel into the eastern parts of the world, to obtain books of the languages for him."[30] Another source is a work by Thomas Smith, published in 1707.

> Mr. Greaves furnished himself with quadrants and other instruments necessary for taking the altitudes and distances of the stars, and the latitudes of cities, for measuring the pyramids, and making observations of the eclipses, at his own expense, having in vain applied for the patronage and assistance of the magistrates of the city of London whose honor and advantage he designed to consult in this voyage; but that he was probably assisted by the archbishop [Laud], who gave him letters of recommendation to Sir Peter Wyche, the ambassador from King Charles I to the Porte, and full power to purchase, at whatever price he thought proper, any manuscripts of value, especially in the Arabic language.[31]

Primordial sources of astronomical language were gradually linked to a larger context in which findings in philology served as authorities in European controversies. The funding for Greaves's trip came from theologians who were concerned about current controversies in political theology. The meeting between Laud and Greaves provided an intellectual and political context for his travels that was connected to a patronage chain—Greaves, archbishops Laud and Ussher, and their ultimate patrons, James I and, later, Charles I.

Politics and the Purification of Language: The Intellectual Culture of James I

The aspiration to solve problems in the language of astronomy involved a larger intellectual culture of early seventeenth-century England. James I (James VI of Scotland) was much concerned for Christendom, whose unity had been shattered by the Reformation and Counter-Reformation and whose peace was constantly threatened and violated. This boded ill for his own kingdom as well as for the other states of Europe. Pleading the cause of independence from rival ecclesiastical jurisdictions and for the exercise of a central power, James labored hard to reconcile the religious extremes of Catholics, Puritans, and Anabaptists. Neither a middle ground nor a request that each side make concessions could achieve this reconciliation. Instead, he invoked the most ancient and fundamental common ground that precedes all splits and deviations—the most primordial, thus most authoritative, ground.

In reforming the Anglican Church, James demonstrated his hopes. He opposed innovations in the worship service by dint of inner conviction, favoring instead the practices and doctrines of primitive Christianity: "whatever has been received from ancient time in the Church, and confirmed by the authority of the divine word, these things we think ought to be preserved and observed most religiously." He proposed a service "common and uniform in all things, not thoroughly defiled by the corruptions of men."[32] He addressed the various calls for a new English translation of the Bible. While Protestants translated the Bible from Latin to different vernaculars, James turned to antiquity as a common ground. For him, the new Bible should be "constant to the original Greek and Hebrew and set forth without note, for that some of them [other translations] enforce a sense further then the text will bear."[33] The king called for a council "whence it would be clear in the case of each doctrine what would be agreeable with antiquity, to the first and purer times of the Christian Church, [and] what was born from and sprang from the inventions of men not long ago."[34]

James had a unique perception of the "council" and closely connected it to the turn toward antiquity. The monarch possessed a divine right to rule, and thus a council should be summoned not by the pope, but, as in the case of the ancient councils, by the emperor. Furthermore, Lancelot Andrews, one of the theologians with whom James was close, added that while the popish council concerned only

Western churches, the historical seven ancient councils included both Western and Eastern churches and embodied the union of Christendom.[35] Thus, for early seventeenth-century royalists, the return to antiquity as a primordial source also implied the unification of East and West.

James and his theologians, then, historicized and localized "antiquity" as a certain space—the Near East—in which one could find historical remains and clues of a pure, ancient wisdom. Moreover, the English Reformed Church appealed to the notion of "ancient Christendom" as the first four centuries preceding the Roman Catholic Church and as the direct primordial source of the English Reformed Church. Thus, James's effort to unify the churches meant an invocation of the early stage of Christianity, identified with the Greek Orthodox Church: "the first mother of all Churches of Christ."[36] The first step would be a reunion with this church.

The Anglican Church was not necessarily a new religious body, but, allegedly, was a direct descendant of those first four centuries of Christianity in the Near East. This political theology aimed at rejecting Catholic allegations that Anglicans, just like the Protestants of the European continent, were breaking from the ancient tradition of the apostles. Politically, the appeal to the Greek Orthodox Church as a role model was made because the head of the Byzantine Empire was also the head of the Greek Orthodox Church, and James tried to shape good relations. Moreover, for international affairs, such a theological statement challenged Cardinal Richelieu's program of converting the Orthodox Greeks of the Ottoman and Safavid empires to Catholicism. While Richelieu used his alliance with the Ottomans against the Habsburgs to gain rights to the working of the Capuchins and Jesuits in the East,[37] James looked to form an alliance with the Greek Orthodox Church based on mutual recognition, not conversion. The hoped-for outcome was a union of the Anglican and Greek churches.

James worked on establishing a Greek printing press in Constantinople, in part to counteract the work of the Medicis and Richelieu, who were printing Arabic-language bibles on their presses to promote Catholic conversions, and later were printing Arabic works of natural philosophy. Also, to balance out the support given by Richelieu and the Medicis to networks of missionaries in the Near East, James initiated an "international student program" in Oxford that admitted and supported Greek Orthodox students from the Ottoman Empire. The most famous among them were Cyril Lucaris and Metrophanes Kristopoulos, who received the king's fellowships. The two studied and exchanged knowledge with Oxford scholars, among whom were Henry Briggs and John Bainbridge, Savilian professors of geometry and astronomy, respectively, and senior colleagues of Greaves.[38] The importation of Greek Orthodox students eventually helped inspire interest in the Near East, and at the same time educated the future leaders of the Greek Orthodox Church. On return to their homeland, Lucaris and Kristopoulos rose in position to patriarchs of Constantinople and Alexandria, respectively, and kept strong connections with English envoys—such as, eventually, Greaves.

James's harmonious worldview would eventually apply to controversies in astronomy. On March 20, 1590, as James VI of Scotland, he had visited Tycho Brahe and his Uraniborg Observatory on the island of Hveen. There, James embraced and praised Tycho's system, which he found to support his Anglicanism—a reconciliation between the traditional Ptolemaic cosmology and the new heliocentric Copernican cosmology.[39] Also, astronomers turned to James. Kepler, fascinated with harmony, found James's idea of unification to be an exemplar for his laws of the harmony of planetary motion. In 1607, Kepler sent James a copy of *De stella in pede serpentarii* and in an attached letter expressed his admiration for the king's learning and conciliatory approach to life. Just as Kepler had brought harmony to the irregularities in astronomy, he wished that God would grant the king power to effect "the pacification and improvement of the church reborn under most difficult circumstances to the well-being of Christendom and safety of the realms entrusted to him."[40] In 1619, Kepler dedicated *Harmonices mundi* to James, appreciative that the king was seeking "harmony and unity in the ecclesiastical and political spheres."[41]

Archbishop Laud and the Authority of Antiquity

Laud and Ussher played a significant part in shaping the political theology of James I and, later, his son, Charles I. In addition, they maintained patron-client relationships with a variety of educated men in England. One of them, Greaves, appropriated their political worldview and derived his methodology in natural philosophy from their treatment of Scriptures.

Laud acted within the close circles of James I, especially as teacher of the successor, Charles I. He took a radical position among the so-called Conformists in the religious controversies, men who believed in a king-led theocracy. Laud did not seek a pleasant consensus, however, as did the Conformists, but proposed superimposing an external order—namely, the ancient church's social structure, political linkages, liturgies, and texts. After the death of James I in 1625, Charles I followed the radical approach proposed by Laud and collided with parliament. In 1626, in a sermon before Charles and the second parliament, Laud addressed the question of political and theological controversies, arguing that the relations between king and church, as well as king and parliament, should rest on the biblical model of the "house of David." Religion and politics were unified in the Temple of Jerusalem, because any contending factions would be subject to the unifying power of "the original copy of the law, the word of God written in tablets of stone."[42] For Laud, a strong relation between politics and religion was embodied in the Old Testament model of David, who derived his political legitimacy from the fact that he represented Mosaic law in the sanctuary. In this way, the king's political authority protected and unified religion, and the king became head of both state and church. As we shall see later, Laud was not the only one who attributed great symbolic value to the Temple in Jerusalem as a microcosmic unification of politics, theology, and law and as an ideal form of authority.

Archbishop Laud focused on something of an epistemological problem: "When matters fundamental in the faith come to question, they finally rest upon a higher and clearer certainty than can be found in either number or weight of men."[43] Neither the pope, as Catholics would argue, nor an individual reading of the Bible, as Protestants would suggest, could express a final judgment on controversy. For Laud, the arbiter was not just the Bible, as other Protestants would argue, but mainly the original versions of the Bible, written in Hebrew, Aramaic, and Syriac, and the early fathers of Christianity. In a debate with the Jesuit Fisher, Laud directly and indirectly attacked Cardinal Bellarmine for his arguments against King James's reforms: "I have expressly declared, that the Scripture, interpreted by the Primitive Church, and a lawful and free General Council determining according to these, is judge of controversies, and no privet [*sic*] man whatsoever is or can be judge of these."[44] Thus, Laud's authority for the resolution of controversies could rely only on the Scriptures of the "Primitive Church." But, how could someone recapture these versions of the Scriptures? For Laud, it began with trying to obtain scriptural versions in Near Eastern languages.

Laud took James's assertions about the Near East a step beyond theory and put them into practice. As a leader at Oxford, he came into close contact with the Greek Orthodox students, who prompted his interest in acquiring deeper oriental sources and a skilled faculty that could read and recapture the primitive sources of the Eastern Church.[45] During his term as chancellor at Oxford, Laud employed Greaves and others as collectors of ancient Near Eastern manuscripts. For instance, he was interested in an Arabic version of the early church councils brought to him by Thomas Roe, the returning English ambassador to Constantinople, and by Cyril Lucaris, one of the Greek Orthodox students.[46] Eventually, Laud enriched the Bodleian Library with such collections of Near Eastern works.

The practice of book collecting was so extensive that Laud instructed Levant Company ships not to return to London without books.[47]

Greaves was one of Laud's messengers to the Near East. Their correspondence frequently performed the necessities of any patron-client relationship. Greaves would update Laud about his findings of books and coins and his work to establish astronomical observations; he even discussed political and cultural aspects of the Near East. But Greaves was not only collecting, he was also delivering, especially items to Patriarch Lucaris (the former Oxford student).[48] He kept Laud informed of the situation in the East. Lucaris continued his efforts to help the English envoys until the bitter end. In his letter of August 2, 1638, Greaves describes the tragic death of Patriarch Lucaris, who was put on trial and executed by the Ottoman government for treason. He was accused of having affinities with foreign political elements and of selling stores of manuscripts from local monasteries to European envoys. Greaves himself was in danger "for having procured out of a blind and ignorant monastery, which depends upon the Patriarch, fourteen good manuscripts of the fathers."[49]

Greaves's relations with Archbishop Laud were based mainly on discussions about the quest for ancient church writings, and less about the practices of astronomy. However, with his other patron—Archbishop James Ussher, who was a well-rounded scholar—Greaves found more in common with regard to the content of his astronomical work and the collection of ancient Arabic and Persian manuscripts of natural history.

Archbishop Ussher and the Unification of Ancient Time and Space

The different approaches of Laud and Ussher to ancient sources paralleled the different practices Greaves applied on his trip. Laud stressed the fixity of ancient sources, whereas Ussher underscored the multiplicity of ancient sources, seen as deviations from one primordial source. Ussher and Laud had a close relationship and corresponded during the years 1628–40. They found common ground in the quest for antiquity, in their royalist zeal, and in their opposition to Rome. However, instead of Laud's creation of uniformity through an "external order of antiquity," Archbishop Ussher saw James I's intellectual culture of reconciliation in a slightly different way. Ussher was interested not just in applying the external authority of the ancient church, but in bridging the gap between the present and antiquity. He did not believe there was a clear cut-off between the two; he thought that certain truths were intertwined in English Christianity, from the "Primitive Church" of the first four centuries up to the Anglican Reformed Church.

One of Ussher's important projects was to show that the Anglican Church was not merely reformist. Instead of using the authority of the ancient church as a substitute for Roman Catholic authority, he built upon latent ingredients of antiquity that had been present throughout the history of the Christian church. His first work, *De Christianrum ecclesiarum successione et statu*, was designed to carry on the argument of John Jewel's *Apologia* (1562),[50] which vindicated Anglican doctrine as the doctrine of the first four centuries of Christianity. Ussher undertook to show a continuity of the same doctrine up to 1513.[51] For the task of historicizing these threads from antiquity to the sixteenth century, he labored to make a biblical chronology (which was later attached to the margins of the King James Bible) and marked the year 4004 BC as the year of the creation of the world (figure 23). The project of biblical chronology actually served Ussher's grander scheme—to count back precisely to the year of the apocalypse, to the end of time.

Ussher's method emphasized mistakes in translation, transcription, and numbers. Mistakes stood as indications of a corruption of antique knowledge and provided threads to follow. Cumulative error, transmitted through translations, eventually broke the harmony and unity of ancient knowledge and created controversy that "arises either through error in men's judgments or else disorder in their affections."[52] In his essay *Tractatus de controversiis pontificiis*, Ussher posited two important insights: first, the "holy spirit expresses itself in our words," and second, "the supreme judge of controversies are [*sic*] the words asserted in the scriptures of the prophets and apostles."[53] The holy spirit thus expressed itself for the first time in antiquity in one language, long before the original and true expression was transmitted and translated into many languages and cultures. Consequently, for Ussher, religious controversies emerged from loss of the true meanings of words, and for the sake of religious peace, it was necessary to restore antiquity's truths.

The actual use of such a technique presented a methodological problem. On the one hand, Ussher believed that the ancient Scriptures should resolve controversies; on the other hand, he was aware of the many different versions of ancient timekeeping and geography. To follow his creative solution, we should look at his textual restoration of ancient Holy Scriptures and his use of astronomy, which ultimately would lead to a determination of an absolute time and place for events of antiquity.

Ussher's restoration of text involved retrieving ancient versions of the Bible, collating them, and finding the most common meanings of words, which then became "true meanings." He was involved in the publication of the Polyglot Bible

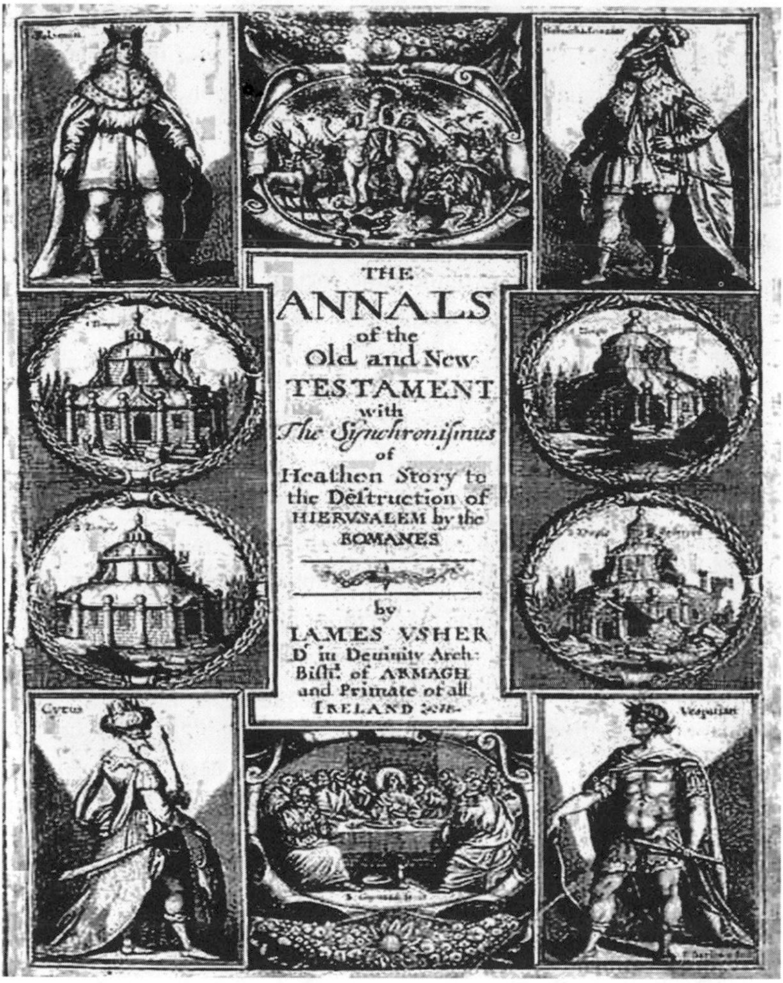
THE
ANNALS
of the
Old and New
TESTAMENT
with
The Synchronismus
of
Heathen Story to
the Destruction of
HIERVSALEM by the
ROMANES

by
IAMES VSHER
D^r in Divinity Arch:
Bish. of ARMAGH
and Primate of all
IRELAND

Figure 23. Title page of James Ussher's *Annals of the Old and New Testament* (London, 1654). Courtesy of The William Andrews Clark Memorial Library, University of California, Los Angeles.

of London in 1647: he collected ancient versions, compared texts, edited them, and sponsored the project.[54] Among other things, Ussher was preoccupied with one of the ancient versions of the Bible—the Samaritan Pentateuch. Thomas Davis, an English merchant working in Aleppo, sent a letter on January 16, 1626, telling Ussher about the successful recovery of ancient sources, including "the five Books of Moses in the Samaritan character."[55]

Ussher made use of the Samaritan Pentateuch to determine the genealogy of the patriarchs and to measure the age of the world since the creation. He transcribed the parts of Genesis 5 and 11 that contain the genealogies. In a letter to John Selden, Ussher compared the different versions of the genealogies from Adam to Abraham as they appear in the Samaritan and Hebrew Pentateuchs and made "a collection of all the differences betwixt the text of the Jews and Samaritan throughout the whole Pentateuch a work which would very greedily be sought for by the learned."[56] Ussher presented a comparison of the different versions (Greek, Hebrew, Samaritan, Syriac, and an Arabic manuscript of the fathers) describing the length of time that passed between the creation and the biblical Flood. However, because the various sources supplied different dates, differing by hundreds of years, he assumed mistakes were made in transcription. To resolve them, Ussher looked for a standard and uniform calendar with which one could make precise calculations of biblical chronology. The linguistic practice he followed reminds us of the bilingual dictionary of Ussher's protégé Greaves and the latter's opinion about the role of intermediary languages.

Ussher considered astronomical events to resolve the disagreements in chronologies and the technical means of timekeeping. A Hebraist colleague, Ralph Skinner, had already addressed some of Ussher's questions in using medieval Hebrew commentaries. In a letter to Ussher in January 1625, Skinner referred to the commentary of Ibn-'Ezra on a lunar eclipse mentioned in the Book of Daniel, and quoted Ibn-'Ezra directly (in Hebrew) as follows: "in the birth year of the King of Persia there was a lunar eclipse, if you would be able to know the precise moment in which the moon was eclipsed we could go backward and find out the other eclipse of the moon, which would tell us how many years passed between the two."[57]

This comment by Ibn-'Ezra, quoted by Skinner, motivated Ussher and Greaves to establish a network of observers for lunar eclipses in various cities in the Near East. On the basis of his interest in the commentaries on the Book of Daniel, Ussher counted the precise age of the world to predict the end of time. The technique he employed to resolve problems in timekeeping and to study the different calculations of the solar year was to use and compare a variety of Greek and Islamic astronomical accounts, looking for their average points and corroborating them by counting back the cycles of contemporary eclipses. He established a rate of divergence among the accounts,[58] and in reconciling differences among chronologies and calendars, he created a universal timeline for the history of the world.

Ussher realized that he must remeasure precisely the time and space of the Near East, and he used Greaves for this, in two ways. First, following Ibn 'Ezra's suggestion, it seemed possible to find the authoritative chronology by using the most ancient recorded celestial event (in the Book of Daniel, as well as a lunar eclipse in the birth year of the Persian king). By calculating the frequency of lunar eclipses in the latitude where the ancient text was written, one could calculate forward and backward and find a uniform timeline in antiquity. Second, Ussher sent Greaves to the areas where the ancient Scriptures were written. An absolute time of antiquity could not be determined without an absolute space of antiquity, so it was also necessary to determine the precise location of places. This could be done by measuring their latitudes and, by creating networks for observing lunar eclipses, measuring longitudes. Using this information, one could calculate a relatively precise frequency of lunar eclipses and determine units of time down to Ussher's and Greaves's own day.

Thus, Ussher used Greaves's astronomical skills and his observational program in the Near East. In December 1638, Greaves established a network of observers in the ancient cities for a lunar eclipse. He trained and instructed local astronomers to conduct observations according to the parallax method of Tycho Brahe. The purpose of this observational networking was to derive the most precise possible data—data both internally rational and relatable to old astronomical tables. Greaves looked for the most accurate and ancient description of a lunar eclipse (in the Book of Daniel) and tried to determine how many years had passed since this celestial event. These calculations helped Ussher make his famous biblical chronology,[59] and by so doing, he could come up with at least a reasonable estimate of the time for events occurring in places where the Old Testament was written.

In addition to determining absolute time, Ussher also worked to determine precisely the places where Jesus preached and where the seven churches of Asia were born. In his *Geographical and Historical Disquisition Touching the Asia Properly So Called*, Ussher found much perplexity in "the several acceptations of the name Asia . . . [And] in reading as well of the New Testament as of other ecclesiastical and civil histories, I endeavored to try whether, by a fit distinction of places and times, some help might be found for the resolving of those difficulties."[60]

The goal of Greaves's travels, then, was to help Ussher determine an absolute spatial reference for places of antiquity. In "An Account of the Latitude of Constantinople, and Rhodes," first published in 1685 in the *Philosophical Transactions* of the Royal Society, Greaves refers to the work of Ussher "as a key to your Grace's exquisite disquisition, touching *Asia* properly so called."[61] He explains that because he dissented from the ancient and new astronomical tables, which he carefully examined in different languages, he had "alter[ed] the Latitudes, if not Longitudes, of most of the remarkable Cities." Greaves gives a survey of the different numbers used for the latitude of Byzantium (Constantinople) by Greek, Islamic, and contemporary European writers. For him, the problem needed to be reconciled, and "the best way to end the dispute, will be, to give credit concerning the Latitude of *Byzantium*, neither to the Greeks, nor to the Arabians." As a skilled judge of these differences, he offers a new measurement: "I have reason of this assertion, [which] appears by several observations of mine at *Constantinople*, with a brass Sextant of above 4 foot Radius. Where taking, in the Summer Solstice, the Meridian Altitude of the Sun . . . I found the Latitude to be 41 degrees and 6 minutes." Greaves further makes a connection between his findings and the need to remake maps: "all Maps for the North East of *Europe*, and of *Asia* . . . are to be corrected, and consequently all the Cities in *Asia* properly so called, are to be brought more southerly then [*sic*] those of *Ptolemy*, by almost two entire degrees, and then those of the *Arabians*, by almost four."[62]

Greaves excuses the mistakes of Ptolemy and the Arabians: because events occurred far from their locations, they "necessarily have depended either upon relations of Travelers, or observations of Mariners, or upon the Longitude of the day." He praises Tycho for his observations and his interest in observed data from other locations. However, for the Copernican system, Greaves finds its mistakes inexcusable, for two reasons. First, Copernicus built his tables without consulting travelers who could measure and report precise data on latitudes. Greaves writes: "I say no man that has conversed with modern travelers and navigators can be ignorant." Second, Greaves attacks the use of precession and trepidation and the implication that the poles of the earth slowly change and that, as a result, the latitudes and calendars also change: "wherefore to excuse these errors of his [Copernicus's] (or rather others fathered by him) with a greater absurdity, by asserting the Poles of the World since his time to have changed their site, and consequently all Countries their Latitudes."[63]

Greaves mistakenly thought that Copernicus's handling of precession would alter the poles of the earth and hence the latitudes of cities. He must have thought, therefore, that if Copernicus had actually measured the latitudes of cities, he would have found out how defective his system was. But we know that Copernicus did not include a table of coordinates of cities, nor do any of his tables depend on latitudes.

In addition to measuring latitudes and longitudes of ancient cities for Ussher, Greaves also worked on purifying ancient words in chronologies, publishing for this purpose *Epochæ celebriores, astronomicis, chronologicis*. Using the astronomical tables of Ulugh Beg, he reconciled ancient chronologies in Syriac, Greek,

Arabic, and Persian in terms of the Julian calendar.[64] Greaves wrote to Ussher in 1644 that "according to your grace's advice, I have made a Persian lexicon out of such words as I met with in the evangelists, and in the Psalms, and in two or three Arabian and Persian nomenclatures. So that I have a stock of above six thousand words in that language." Then he asked Ussher to support his travel and to ensure his position on his return to Oxford, because his journey would be for "improvement of learning and for the publishing of some of those books. There I shall have the opportunity of printing your grace's map."[65] Following in Ussher's footsteps, Greaves could effect a unity of ancient and present time in making a universal calendar. Thus, Greaves proposed to reform the Julian calendar, still used in England, and to follow the Gregorian reform and unify Europe's and England's timekeeping.[66]

At this point, we can see how social circles, patrons, agendas, and controversies were reflected in Greaves's labors in astronomy. On the one hand, Laud employed Greaves for his interest in obtaining Persian and Arabic manuscripts of the "Primitive Church." On the other hand, from his social setting, and especially from Ussher, Greaves derived his practice of reconciling controversies in astronomy and geography. The process involved collating sources of astronomy, including late fifteenth-century Arabic and Persian versions, and subjecting them to linguistic testing against the authority of, purportedly, the most ancient versions. By so doing, he did not superimpose antiquity on the controversies, but stressed the latent continuity between antiquity and the present. Ussher and Greaves had chosen to look into the roots of contemporary controversies in the fields of language and history. For both men, the differences in accounts of time and place were not a result of a changing nature, but a result of changes in the knowledge of nature or, more accurately, a distortion of the perfect knowledge of nature that had existed in antiquity. These practices were set against the Linguistic Leviathan.

The Wider Oxford Circle

The quest for the Linguistic Leviathan was a reflection of the pressing needs of Greaves's patrons. But was Greaves alone and unusual? Did the controversies inspire other practitioners of natural philosophy to search for primordial sources? He had encountered models of linguistic criticism among some of his predecessors at Gresham College,[67] but we can find the direct models among his associations with other protégés of Laud and Ussher who were Greaves's senior colleagues, men such as Henry Briggs (1561–1630) and John Bainbridge (1582–1643).

Bainbridge preceded Greaves in publishing corrected editions of Arabic astronomical texts. After studying Arabic, he wrote a bilingual Arabic-Latin treatise on astronomy, entitled *Canicularia*, which Greaves published in 1648 after Bainbridge's death.[68] Bainbridge had not worked strictly on his own, and in general he deployed his astronomical and linguistic skills for the service of Ussher's project, especially in calculating dates by means of ancient eclipse records. For instance, Bainbridge answered a question addressed by Ussher on the occurrence of a solar eclipse "*in anno periodi Julianae 4114*," and in a letter dated April 1624 he wrote to Ussher that "after a diligent search I find that none [no solar eclipse] could appear in Europe or the confines of Asia; but in the former year of 4113 . . . According to Ptolemy and his tables in the meridian through Alexandria, Rhodes and the western part of Asia Minor . . . there appeared a notable eclipse of the sun, three hours and twenty-five minutes."[69]

Unlike Greaves, who embarked on travels for Laud and Ussher, Bainbridge did no more than consult astronomical tables in available texts. Bainbridge's *Canicularia* aimed to resolve such controversies by returning to primordial sources and to a purified set of astronomical nomenclature and language.[70] He was a source of inspiration for Greaves, who took his project one step further.

Briggs, by contrast, was unskilled in Eastern languages and his knowledge was mainly in mathematics and cartography,[71] and he appropriated the search for primordial sources into his mathematics. He was fascinated by the invention of logarithms, and even paid a visit to Napier and offered to explain for him the essentials of the ten-base logarithm. Logarithms, for Briggs, were loan-numbers that were joined proportionally to their main numbers, and hence "any proportional numbers therefore being given, divers other numbers may be annexed into them, exactly agreeing with the general definition of Logarithms."[72] Briggs perceived logarithms as artificial pyramidal numbers that proportionally represent true numbers—in other words, they are pyramidal representations of primordial numbers. Although Briggs corresponded with Kepler and was well aware of continuing changes in cosmology, he did not take a clear stance supporting the Copernican system.[73]

Bainbridge and Briggs were also engaged in other activities in the service of Conformist clergymen, especially in comparing ancient patristic manuscripts.[74] Thus, the two men not only were working with their skills in natural philosophy, but also labored in the heart of theological disputes. They did not live into the 1640s to experience persecution and loss of position at Oxford, as did other members of the circle.

Laud employed yet another close friend of Greaves, not to help in research into natural philosophy or political theology, but for reforming the administrative system at the University of Oxford. By the order of Laud, Peter Turner (1586–1652), who introduced Greaves to Laud and who succeeded Henry Briggs as professor of geometry at Gresham College, served on a committee to revise university statutes and "to reduce them to a better form and order."[75] After the death of Briggs in 1630, Turner succeeded him as Savilian Professor of Geometry at Oxford. On his appointment as chancellor of the university in 1631, Laud urged forward the work of administrative reform. Final revision of the university statutes was entrusted to Turner, who was requested by Laud "to polish the style and prepare it for the press."[76] This book of statutes was published in 1634. Some of the notions in the preface of the statutes concerned the type of external order needed to solve religious crises. During the Civil War, Turner was taken prisoner, and in 1643 he was exchanged for some parliamentarian prisoners at Oxford. In November 1648, the parliamentarian commissioners ejected him (at the same time as his friend Greaves) from his fellowship and Savilian professorship.

Another protégé of Laud from Oxford, one who accompanied Greaves on his travel to the Near East, was Edward Pocock (1604–91), who first traveled to Aleppo between 1630 and 1635. On that earlier visit he had mastered Arabic, Hebrew, Samaritan, Syriac, and Ethiopic, and had associated with learned Muslims and Jews who helped him in collecting manuscripts. In his travels, he collected a large number of such texts, including a Samaritan Pentateuch. All of this served Laud's project well, and Pocock eventually attracted his notice. Laud wrote to him several times with commissions for the purchase of ancient Greek coins and oriental manuscripts. After becoming the archbishop of Canterbury and chancellor of the University of Oxford, Laud offered to appoint Pocock as the first professor of Arabic. Pocock returned to England, probably early in 1636, and took up this position. However, in 1637, at Laud's request, Pocock again set sail for the purpose of further research into manuscripts. Now he traveled with his "dear friend" John Greaves. The two men divided the mission into two tasks: Pocock would collect manuscripts and Greaves would conduct celestial observations and measurements of ancient monuments. During the Civil War, Pocock fared better than his colleagues. Although he visited Laud in the Tower of London and for a time lost his position at Oxford, he regained his position based on his close relationships with scholars who were parliamentary sympathizers, including Robert Boyle.[77]

In summary, then, archbishops Laud and Ussher employed scholars in Oxford, including Briggs, Bainbridge, Turner, Greaves, and Pocock, to recover ancient texts and to resolve controversies in political theology by pointing to ancient authorities. The scholars, in turn, internalized this methodology and searched for ancient textual authorities to resolve controversies in astronomy.

Copernican Parliamentarians

James Ussher consulted with several scholars on the matter of observing eclipses and their use in establishing ancient chronology. Among them was William Gilbert (1597–1640), an enigmatic figure,[78] who wrote to Ussher on December 11, 1638, about their plan to observe an eclipse and, in passing, inserted a Copernican argument: "All my expectancies for observation of this lunar eclipse last Tuesday morning, were lost in the cloudy disposition of the heavens for that time." Then he tried to convince Ussher to apply principles of Copernican cosmology to the project of construction of "universal time," because these principles "excellently accommodate many irregular motions to account, and open a large field for the search and invention of high things."[79] Besides presenting a new heliocentric cosmology and new techniques for measuring the motion of the celestial bodies, Gilbert also advised Ussher that his obsession with the date and workings of the creation of the world were irrelevant if humanity were not the center of creation and the center of God's attention.

There were other, more prominent Copernicans. John Wilkins (1614–72), for instance, was first a silent and later a vocal supporter of the new cosmology. In 1638, during Greaves's excursion, Wilkins published anonymously his first work, *The Discovery of a World in the Moone* (figure 24). The book contained a diagram of the heliocentric Copernican universe, and Wilkins tried to prove, by relying on Galileo's discoveries, that the moon had valleys and mountains, lakes and seas, and therefore was a habitable world.[80]

Another of Wilkins's works, published in 1640, advocated the Copernican system; it was entitled *A Discourse Concerning a New Planet* and demonstrated that Earth is but one of the planets.[81] But more significant to our discussion is the epistemological device for supporting the argument for life on the moon. Wilkins writes: "the new truth may seem absurd and impossible not only to the vulgar, but also . . . to scholars . . . and hence it will follow, that every new thing which seems to oppose common Principles is not presently to be rejected, but rather to be pried into with a diligent enquiry, since there are many things which are yet hid from us, and reserved for further discovery."[82]

Wilkins was a parliamentarian and a central figure in England's Civil War, and he appreciated new structures. He stressed that scholars should not be deterred by new views: some common opinions could eventually come to seem absurd

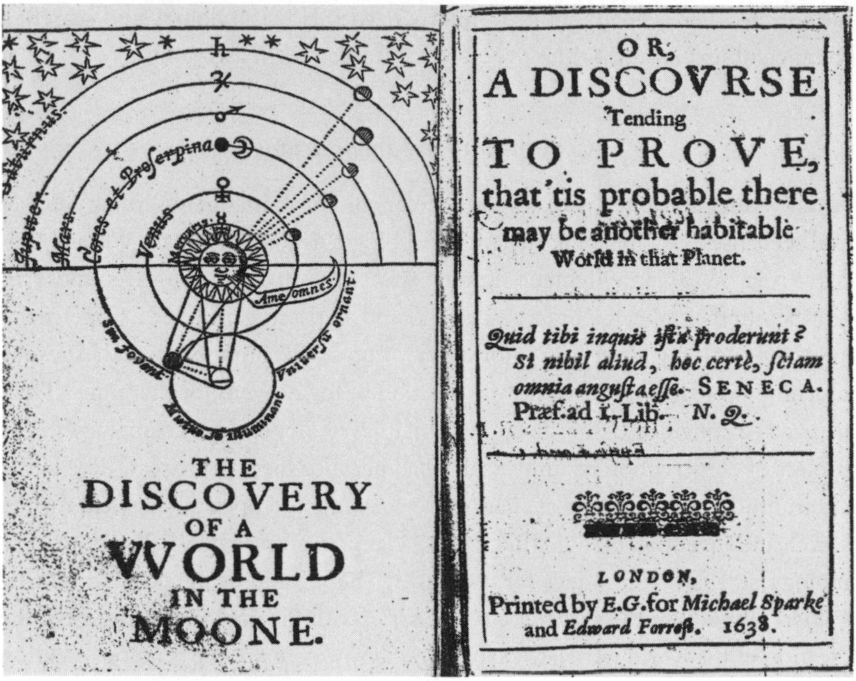

Figure 24. Title page from John Wilkins's work on "another habitable world" on the moon. From Wilkins, *The Discovery of a World in the Moone* (1638), QB41. Courtesy of The William Andrews Clark Memorial Library, University of California, Los Angeles.

("what the discovery of the new world did to geography"), and new findings that were criticized could later be proved as truth (such as the Copernican system). Wilkins, simply speaking, believed that all things were possible. He stated that he would confirm his view about a world on the moon "by sufficient authority of diverse authors, both ancient and modern, that so I may the better clear it from the prejudice either of an upstart fancy, or an absolute error."[83]

The adversarial relationship between Greaves and Wilkins went beyond cosmological controversy and touched their careers as professors at Oxford. During the Puritan rule of the 1640s, visitors from parliament began a university reform. They wished to clean the university of traces of Laud and his followers. First, they ejected the Laudian circle, who were replaced by close colleagues of Wilkins: John Wallis replaced Peter Turner, and Seth Ward replaced Greaves, as Savilian professors of geometry and astronomy, respectively; Wallis and Ward would later be involved in controversies with Hobbes.[84] Wilkins was then responsible for deciding whether Greaves was to be compensated for his loss.[85] The shift at Oxford from royalist to parliamentarian sympathies was also a shift in intellectual culture from a textually inclined deductive methodology to an experimental-inductive exploration of nature.

From the "Linguistic Leviathan" to the "Nature of Things"

The struggle between the two camps extended beyond professional intrigues, political theology, and cosmology, and was reflected in notions about a possible universal language for natural philosophy. While Greaves and his colleagues hoped to recapture language through material ancient objects, Wilkins was interested in creating language for use in natural philosophy. In 1668, Wilkins published *An Essay towards a Real Character and a Philosophical Language* (figure 25), which considered the barriers to communication caused by the diversity of languages and their written symbols. Instead of attempts, as in Greaves's case, to look for universal language "from [a] Dictionary of Words, according to some particular language, without reference to the *nature of things*,"[86] Wilkins suggested the invention of a new universal writing system based on a set of devised characters. These could express things and notions recognizable to all people at all times.

> The several Nations of the World do not more differ in their Languages, then [*sic*] in the various kinds and proportions of these Measures. And it is not without great difficulty that the Measures observed by all those different Nations who traffic together, are reduced to that which is commonly known and received by any one of them; which labour would be much abbreviated, if they were all of them fixed to any one certain Standard. To which proposed, it were most desirable to find out some *natural Standard*, or *universal Measure*, which hath been esteemed by Learned men as one of the *desiderata* in Philosophy. If this could be done in Longitude, the other Measures might be easily fixed from thence.[87]

And, indeed, the measurement of longitudes with clocks was the source of inspiration for Wilkins's proposed mechanical experiment to measure length by using time. He offered to build a pendulum so "that the space of every Vibration be equal to a second Minute of time; the String being, by frequent trails, either lengthened or shortened, till it attain to this equality." The vibrations should last a sufficient time, either five or six hundred vibrations, and pass through an arc

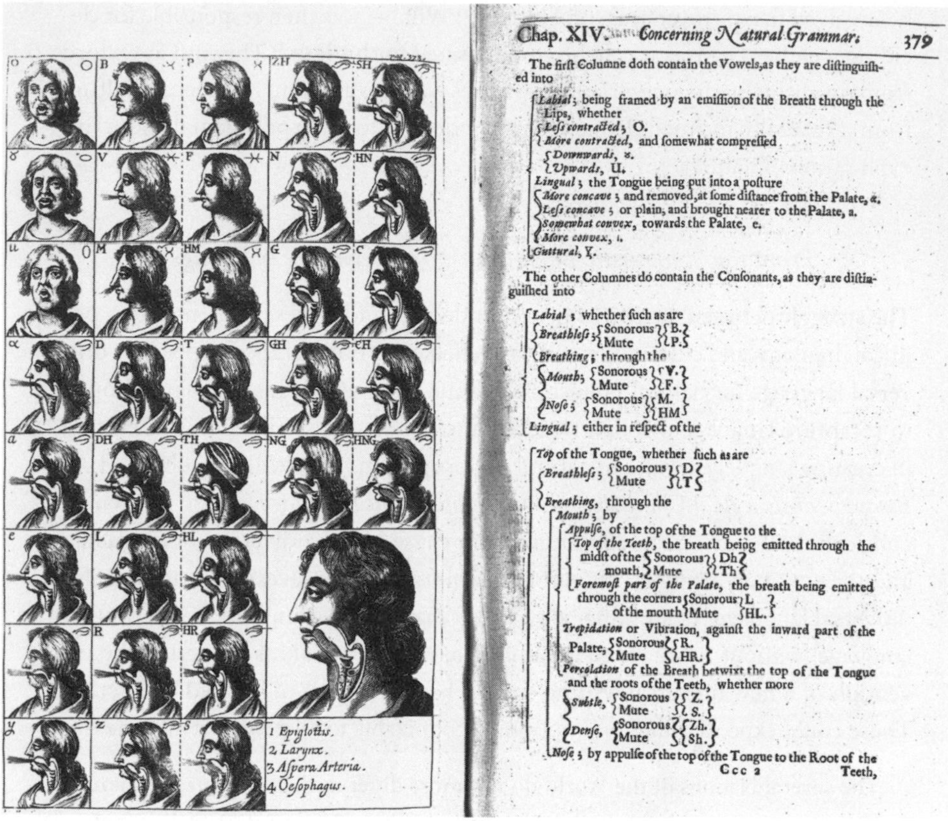

Chap. XIV. *Concerning Natural Grammar.* 379

The first Columne doth contain the Vowels, as they are distinguished into

- *Labial*; being framed by an emission of the Breath through the Lips, whether
  - *Less contracted*; O.
  - *More contracted*, and somewhat compressed
    - *Downwards*, ȣ.
    - *Upwards*, U.
- *Lingual*; the Tongue being put into a posture
  - *More concave*; and removed, at some distance from the Palate, α.
  - *Less concave*; or plain, and brought nearer to the Palate, a.
  - *Somewhat convex*, towards the Palate, e.
  - *More convex*, ı.
- *Guttural*, Y.

The other Columnes do contain the Consonants, as they are distinguished into

- *Labial*; whether such as are
  - *Breathless*; Sonorous B. / Mute P.
  - *Breathing*; through the
    - *Mouth*; Sonorous V. / Mute F.
    - *Nose*; Sonorous M. / Mute HM
- *Lingual*; either in respect of the
  - *Top* of the Tongue, whether such as are
    - *Breathless*; Sonorous D / Mute T
    - *Breathing*, through the
      - *Mouth*; by
        - *Appulse*, of the top of the Tongue to the
          - *Top of the Teeth*, the breath being emitted through the midst of the mouth, Sonorous Dh / Mute Th
          - *Foremost part of the Palate*, the breath being emitted through the corners of the mouth Sonorous L / Mute HL.
        - *Trepidation* or Vibration, against the inward part of the Palate, Sonorous R. / Mute HR.
        - *Percolation* of the Breath betwixt the top of the Tongue and the roots of the Teeth, whether more
          - *Subtle*, Sonorous Z. / Mute S.
          - *Dense*, Sonorous Zh. / Mute Sh.
      - *Nose*; by appulse of the top of the Tongue to the Root of the

Ccc 2 Teeth,

Figure 25. Page from John Wilkins's *Essay* showing an embodiment of natural properties as constituting a universal language. From Wilkins, *An Essay towards a Real Character, and a Philosophical Language* (1668), P101. Courtesy of The William Andrews Clark Memorial Library, University of California, Los Angeles.

of five or six degrees, and then one would "measure the length of this String . . . [and that] being done there are given these two Lengths, *viz.* of the *String* and of the Radius of the Ball, to which a third proportional must be found out. Which being so found let two fifths of this third Proportional be set off from the Center downwards, and that will be the measured desired (39 inches and a quarter) . . . Let this Length therefore be called the Standard; let one Tenth of it be called a foot; one Tenth of a Foot, an Inch."[88]

Instead of using primordial sources of language as a basis for universal language, Wilkins wanted to rely on the universality of nature as a source for a new universal language for natural philosophy. Just as time is measured by motion,

Wilkins suggested measuring dimension by time, and then there would be no need for historical measurements. Thus, according to Wilkins's experimental and mechanical view, a universal language of natural philosophy would be based on *natural things*, with no need for recapturing ancient knowledge from historical Hebrew, Persian, and Arabic texts. Whereas Greaves was thinking about natural philosophy within the context of history, Wilkins was thinking outside history. As a result, in Wilkins's scheme, the Near East was overlooked.

Although, Laud's and Ussher's use of Greaves's research for their own ends essentially had conservative implications, it is hard to divide the pre–Civil War academic community into radicals and conservatives simply on the basis of their attitudes toward a hermeneutic of ancient texts, on the one hand, and experimentalism, on the other. Briggs, for example, was certainly not opposed to experimentalism. His public support for the Northwest Passage voyages seems to suggest a different spirit than that of Greaves. However, the drive to discover never-before-seen territories is, in a way, the mirror image of the effort to recover a pristine past in ancient lands. Amir Alexander suggests that for mathematicians like Briggs, the quest for a hidden "golden land," by overcoming formidable obstacles, became a metaphor for the quest for knowledge in general and mathematics in particular.[89]

The Reception of Greaves's Works

The project to recover the Linguistic Leviathan did not produce results during Greaves's lifetime, and Thomas Birch, who published the *Miscellaneous Works of Mr. John Greaves* in 1737, stressed that the aim of the publication was "to rescue the writings of great men from obscurity."[90] In modern scholarship, especially on Orientalism, Greaves is mentioned as an eccentric professor of astronomy who had a great interest in the pyramids and in Arabic and Persian astronomical manuscripts.[91] Other accounts mention him both as one of the first orientalists to appreciate Arab-Persian culture and as a scholar who suggested a reliance on fifteenth- to sixteenth-century Islamic astronomy—even after the rise of the new astronomy in Europe.[92] However, few have looked into the question of how his project was received, especially in the light of the close reading of his works by some significant later scholars.

In *A Dissertation upon the Sacred Cubit of the Jews*, first published in 1737 in the *Miscellaneous Works of Mr. John Greaves*, Isaac Newton summarized the accounts on the "Roman foot" and the "Pyramids." Newton writes that Greaves stimulated his fascination with the Jewish Temple, because "to the description of the Temple belongs the knowledge of the Sacred Cubit."[93] Newton, just like Greaves, looked for a primordial source, the Temple's sacred cubit, which acted, in a sense, as a historical a priori that could reconcile the differences among the many ancient accounts of measurements.

But Newton took this description a step further. He suggested that in antiquity, there were two measures of the cubit: the vulgar and the sacred. In exploring the differences among the ancient vulgar cubits, Newton dismissed the Arabian measurements of the mile and cubit, because Arabians "learned from the conquered people the money, weights, and measures of the Romans and the Greeks. We should pass over this Cubit and proceed to those which are more ancient." Newton goes on to describe early measurements of the vulgar cubit. Relying on Greaves's accounts, he asserts that differences cropped up and the vulgar cubit increased in length because of changing instruments of measurement or different sizes of the human body: "the instruments, which were preserved as standards of measures, by contracting rust are increased. Iron beaten by the hammer may insensibly relax in a long space of time." And differences became evident "by comparing the Feet and Cubits used at first in every nation according to the proportion of the members of a man, from which they were taken. For the Foot of a man is to the Cubit of lower part of the arm of the same man."[94]

For Newton, the size of the sacred cubit was the primordial source for all the vulgar cubits, and this measure was spread by the Jews to other nations. Noah was the first to use this cubit in building his ark, but from Noah onward, "knowledge had fallen" and only the Jews continued to use the sacred cubit with which Moses had built the tabernacle and Solomon had built the First Temple. "The *Jews*," Newton writes, "when they passed out of *Chaldea*, carried with them into Syria the Cubit which they had received from their ancestors. This is confirmed . . . by the dimensions of Noah's Ark . . . [Therefore, all the ancient Cubits] derived in different countries from the same primitive Cubit . . . [But when the Jews] afterwards going down into Ægyptians, and enduring an hard service under them, especially in building, where the measures came daily under consideration; they must necessarily learn the Ægyptian cubit." From this arose "the double Cubit of the Jews,"[95] the sacred and the vulgar.

In recapturing the sacred cubit, Newton adopted the grand scheme of Greaves and divided human knowledge into two kinds: one divine and sacred, the other human and vulgar. Unlike Wilkins and his natural science experiments and measurements, Newton thought historically and believed that universal measurements were deducible from antiquity. These notions coexisted with his belief that a universal explanation for nature was lost in antiquity.[96]

We have traced here the social and intellectual threads of John Greaves's work and method, looking at his travel to the Near East and then back to the political, theological, and cosmological questions raised within a circle of leaders at Oxford. The most important finding is that his projects were by-products of the larger projects of his patrons, archbishops Laud and Ussher, who employed the results of Greaves's skills in various ways in their discussions and responses to changing political and philosophical necessities.

In the early seventeenth century, James I of England advocated a bold resolution of the struggle between Catholics and Puritans, one that would help him maintain political power. Assisted by a circle of theologians, he stressed that the Anglican Church was a direct continuation of the "Primitive Church" that took shape in the first four centuries of the Christian church in the Near East. Thus, the resolution of seventeenth-century controversies required a return to the most ancient of accepted authority and a recovery of primordial sources of language and theology.

The approach to conciliation generated research and speculation on the sources of the Anglican Church and on ancient Near Eastern languages and writings, which in turn spoke to problems in astronomy. After the death of James I, Charles I, assisted by William Laud, ventured a radical shift: he wished to superimpose on the Anglican Church a reading of the Scripture of the Primitive Church, with its implication of the perfect union of politics and religious truth within the Temple. As a result, there was the need to find the texts of the Primitive Church.

With the patronage of archbishops Laud and Ussher, Greaves was in effect paving his way to an academic career as professor of astronomy at Oxford. He fashioned his passion for primordial sources in astronomy to fit the needs of his patrons for the primordial sources of the Primitive Church. His attempt to purify the technical terms of ancient astronomy had led him to his work in precision measurement and metrology. He then appropriated his patrons' search for Primitive Church sources into his own exploration of absolute numbers and, finally, the idea of universal language. His close colleagues had the same inclination. They either followed the royalist method of religious unification in writing their own works, or they labored directly in the service of Laud and Ussher, traveling to recover ancient time and space. During the Civil War, the royalist circle lost academic positions at Oxford to a new Puritan and parliamentarian circle led by John Wilkins. Unlike the Tychonic astronomers who were searching for primordial sources of universal authority, the new parliamentarian circle advocated the Copernican system and searched to establish a universal language for natural philosophy based on the universality of nature. Later, Isaac Newton adopted his grand scheme and picked up the work on ancient measurements in an attempt to recapture the measures used in constructing the ancient Jewish Temple, which he considered to be a microcosm of the universe.

These findings, in turn, meld the questions of English Orientalism, experimentalism, and precision. We learn that Greaves was not merely interested in recapturing astronomical texts in Arabic and Persian, but was looking to reconcile contemporary controversies in astronomy by castigating, purifying, and standardizing the language and numbers of astronomy. We also learn that the debate in seventeenth-century England on the nature of the scientific community and its trustworthiness was occurring not just between those who chose to argue from rational first principles and those who supported experimentalism. There was another trend: natural philosophers who distinctively conceived in historical terms the transmission and development of knowledge about nature. They constructed natural philosophy by searching for first principles, as found in primordial languages and numbers, through travels in the Near East. Such scholars aimed to recover the "Linguistic Leviathan," to establish new "values of precision" from ancient numbers and words that could set up an objective and responsible public language and would enable natural philosophers and theologians to reconcile the controversies.

Exchanging Heavens and Hearts

In 1634, Tommaso Campanella joined the court of Cardinal Richelieu in Paris and stimulated there a great interest in the Copernican cosmology. In the same year, a court cosmographer named Noël Duret dedicated to Richelieu a book entitled *Nouvelle théorie des planètes*, which included astronomical tables calculated from the tables of Ptolemy, Copernicus, Tycho, and Lansbergen.[1] In 1660, the Ottoman scholar Ibrāhīm Efendi al-Zigetvari Tezkireci translated Duret's book into an Arabic manuscript.[2]

Early modern scholars came across books in a variety of ways. They may have searched for important books they had heard of or books they had been assigned to read. Sometimes, they randomly surveyed book stacks and accidentally stumbled upon a new text. Of the many such encounters, very few sparked the reader's interest and intrigued him enough to make him want to introduce the book to others. For a scholar living in a scribal culture, the rare encounter with a book from a foreign locale must have created a magical moment in which various streams of cultural consciousness converged. Al-Zigetvari's encounter with Duret's book was one such moment.

By tracing the mechanism of circulation of Duret's book and al-Zigetvari's encounter with it, we can recapture al-Zigetvari's motives and the cultural context in which he deliberated as he made his translation. However, we meet roadblocks in all directions. Who was Noël Duret, and how significant was his book in relation to other books on the Copernican system? His name does not appear in major works, and he is mentioned only in passing as someone who wrote on Kepler. His major work, *Novæ motuum cælestium ephemerides Richelianæ*,[3] published in 1637, deals with astrology, hermeticism, and mysticism and, in passing, mentions the heliocentric Copernican system.[4]

Duret's publications took a convoluted course. He was inspired by the foremost Dutch Copernican, Philips Lansbergen, who emphasized the relation be-

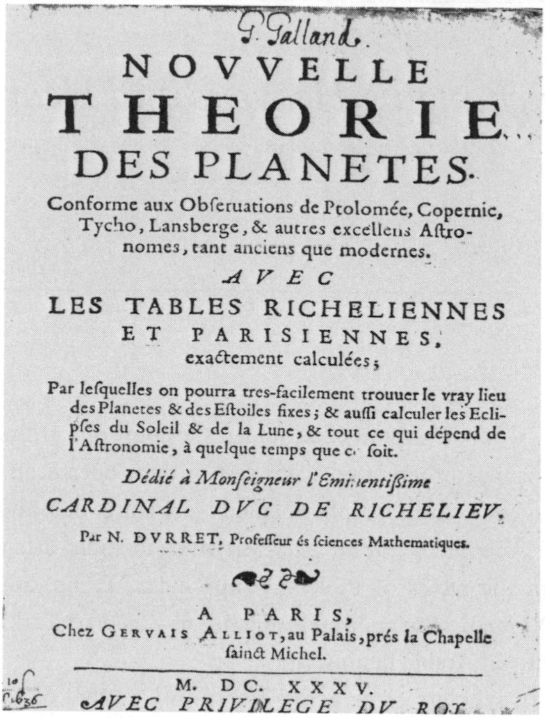
G. Galland.

NOVVELLE

THEORIE

DES PLANETES.

Conforme aux Obſeruations de Ptolomée, Copernic, Tycho, Lansberge, & autres excellens Aſtronomes, tant anciens que modernes.

AVEC

LES TABLES RICHELIENNES ET PARISIENNES, exactement calculées;

Par leſquelles on pourra tres-facilement trouuer le vray lieu des Planetes & des Eſtoiles fixes; & auſſi calculer les Eclipſes du Soleil & de la Lune, & tout ce qui dépend de l'Aſtronomie, à quelque temps que ce ſoit.

Dédié à Monſeigneur l'Eminentiſſime

CARDINAL DVC DE RICHELIEV.

Par N. DVRRET, Profeſſeur és ſciences Mathematiques.

A PARIS,

Chez GERVAIS ALLIOT, au Palais, prés la Chapelle ſainct Michel.

M. DC. XXXV.

AVEC PRIVILEGE DV ROY.

Figure 26. Title page of Noël Duret's *Nouvelle théorie des planètes* (Paris, 1635). Courtesy of Owen Gingerich.

tween celestial and terrestrial mechanics, especially in describing the earth as a ship that was to carry humanity to the heavenly Jerusalem. Following in Lansbergen's footsteps, early in his career Duret published the *Nouvelle théorie* (figure 26), essentially based on Lansbergen's popular tables, which imitated the Alfonsine tables.[5]

In the *Nouvelle théorie,* Duret mentions the Rudolphine tables only in passing, as an example of tables named after a monarch. In 1637 he extended his work and his commitment to Richelieu by publishing the extensive ephemerides, *Novæ motuum cœlestium ephemerides Richelianæ,* dedicated to and named for his patron, Richelieu. Through the Richelian tables, which were based on the tables of Philips Lansbergen and the Rudolphine tables, one could calculate astronomical positions and compare them with current observations or with earlier positions calculated by Ulugh Beg, and could also use the tables to make an almanac. However, when Duret was part way through computing his extensive ephemerides from the earlier tables, he discovered that in 1631, the Lansbergen tables erred

Figure 27. Title page of Noël Duret's *Novæ motuum cælestium ephemerides Richelianæ* (The New Richelian Ephemerides). Note the pillars of astronomy, according to Duret: Lansbergen and Tycho. This page is from the London edition of 1647. The London reissue, which has the same printed pages as the French edition, was given a new title page. The English publisher also bought the remaining pages of Duret's seminal work *Supplementi tabularum Richelienarum*, which were also reissued in London in 1647, again with a new title page. Courtesy of The William Andrews Clark Memorial Library, University of California, Los Angeles.

in predicting the transit of Mercury observed by Pierre Gassendi. Duret then turned his back on his own Richelian tables and switched in midstream to using the Rudolphine tables. In 1639, he published his monumental *Supplementi tabularum Richelienarum, pars prima, cum brevi planetarum, Theoria ex Kepleri, sententia*, in which he very accurately simplified the Rudolphine tables. The publisher's remainders of the *Supplementi* were later sold and reissued in London in 1647 with a new title page[6] (figure 27).

Duret's *Nouvelle théorie* circulated from Paris outward, as a material object that carried textual information on post-Copernican astronomies. At the outer reaches of this circle, in Istanbul, al-Zigetvari encountered the book. Although we find only a few clues on al-Zigetvari's identity,[7] it seems that he borrowed, exchanged, and altered the European astronomical tables and transformed them into ephemerides that were of use to local astrological practices prevailing in his mystical surroundings. The title he gave to the translation echoes medieval sources such as the Marāgha Observatory and men such as al-Ṭūsī (d. 1274) and Quṭb al-Dīn al-Shīrāzī (d. 1311), who had been heavily engaged with the philosophy of illumination, with which al-Zigetvari was affiliated. Thus, al-Zigetvari read, translated, and commented on Duret's book within the context of Sufi practices that strived for illuminist apperception (*idrāk*), which emerges as an illuminist's alternative perception of nature. The incommensurability between nature as perceived through the senses and as understood through reason could be reconciled by *idrāk*, which in hermetic sources was sometimes identified as intuitive perception of angels. For al-Zigetvari, *idrāk* served as a tool for harmonizing the inconsistencies in astronomy. Illuminist intuitive perception was worked out culturally in Sufi practices and in artistic styles and also functioned as an indicator of sincerity and trustworthiness. With *idrāk* in mind, al-Zigetvari circulated new cosmologies, not as breaking from Muslim astronomical tradition, but as encompassing medieval mystical strivings for harmony with the heavens.

Noël Duret and Richelieu

Noël Duret and his work have not survived the centuries well. Duret is mentioned in a letter of July 24, 1675, sent to the Royal Society of London by John Flamsteed, who was appointed by royal warrant as "The King's Astronomical Observator" and became the first "Astronomer Royal." Flamsteed complained in the letter that the "Richileu's Tables . . . are the same with the Rudolphin."[8]

Although Duret and his book were at the margins of contemporary scholarship, we can find some references to them. In 1651, fourteen years after the book's publication in Paris and four years after publication in London, Nicholas Culpeper, an English writer on astrology, medicine, and herbs, published *Semiotica urinica*, a book that dealt mainly with astrology and medicine.[9] Culpeper and his book were widely known among those connected with these two arts, and *Semiotica urinica* circulated in five editions between 1651 and 1671. It set forth two different astrological methods for understanding disease. The first was that of the "Arabian Physician, and Singular astrologer . . . Abraham Avenezra."[10] The second was based on Duret's work, "wherein is laid down the way and manner of finding out the cause, change, and end of the disease: also whether the sick be likely to live or die . . . whereunto is added, a table of logistical logarithms, to find the exact time of the crisis."[11] Culpeper mentions Duret as "Cosmographer to the King of France, and the most excellent Cardinal the Duke of Richelieu,"[12] for whom Duret named and to whom he dedicated *Novæ motuum cælestium ephemerides Richelianæ.* "Henceforth," Duret wrote in the dedication to Richelieu, playing on Lansbergen's metaphor of the earth as a ship, "for all star-lovers who intend to stretch their sails on the deep sea of divine Astronomy, with their anchor now settled in this Parisian port of their own learning, under favorable omens this remarkable volume of ours will shine forth; . . . and the most brilliant star will rise up to guide the ship's oar, . . . by whose brightly-gleaming splendor the black cloud will be broken up."[13]

Nothing in the accounts of Richelieu mentions Duret and the possible transmission of his works to the Ottoman Empire, but one salient fact indirectly links the two men. Richelieu had cultural dealings with the Ottoman Empire. To meet the challenge of France's natural enemies, the Habsburgs, Richelieu continued France's tradition of alliance with the Ottomans.[14] He sent a delegation of diplomats that included humanists, such as Guillaume Postel, who acted not only as diplomatic mediators but as collectors of artistic and scholarly objects. A long tradition of gift exchange existed between the Ottoman court and certain European courts, and the objects exchanged generally related to natural philosophy or art.[15] Rather than being a formal code of conduct, gift-giving was a strategy for success, an external display of power.[16]

Richelieu established famous collections of art and books.[17] He received several scientific and artistic objects from Muslim sources at the beginning of the seventeenth century—one Islamic miniature painting and four samples of manuscripts in Arabic.[18] In all, Richelieu owned thirty-seven manuscripts in Arabic, Ottoman-Turkish, and Persian. François Savary de Breves, who served as ambassador to Constantinople from 1604 to 1613, assembled the collection for Richelieu.[19] On de Breves's return to France, Richelieu promoted him to the post of his personal advisor. From Paris, de Breves maintained close relations with his successor in Constantinople, Harlay de Cèsy, who served as ambassador until 1640.[20] In the process of gift exchange, de Cèsy presented to the Ottoman court examples of the intellectual fruits of Richelieu's court—including Duret's *Nouvelle théorie.*[21] Arabic manuscripts were transferred to Richelieu's library in 1640, and all these exchanges of books took place in the time between the publication of Duret's book in 1637 and the death of Richelieu in 1642. Duret's discussion of the Copernican system seems to have gathered dust in the Ottoman court library for twenty years, until al-Zigetvari came across it in 1660.

If one assumes there are logical agendas in cultural circulation, how did al-

Zigetvari happen to translate a book that, logically, he should not even have encountered? Why did a marginal and rare book become the first reference source on the Copernican system in Arabic, instead of, for example, a work by Copernicus himself or by Tycho Brahe,[22] Galileo, or Kepler? Moreover, al-Zigetvari had to fight to convince the authorities of the utility of translating such a book. Did al-Zigetvari first read Culpeper's book, which was well known in many regions, including the Ottoman Empire, and then, with the endorsement of that author, seek out the work of Duret to translate it? There is no evidence to support such a supposition.

It is useful simply to consider here nonprogrammatic, chance history. Given the nature of gift exchanges, as described above, and the collection of books and art, both in Paris and in Istanbul, it is far more likely that al-Zigetvari did not search for a book on the Copernican system to translate into Arabic. His encounter with Duret's book, then, was a result not of any awareness of the heliocentric system but of mere chance. Once in contact with the book, his approach to the translation—his habits of interpretation and philosophical underpinning—was in great part a product of contemporary cultural trends. These trends were completely suited to advancing knowledge on the techniques and problems, as well as the metaphysics and epistemology, of astronomy.

Al-Zigetvari and His Scholarly World

Ibrāhīm Efendi al-Zigetvari Tezkireci, as we have seen, is a name without links to the standard biographical sources. "Ibrāhīm Efendi" is a common name. A last name normally has a marker of lineage (Ibn or Abu) that might reveal something of a person's origins, and the absence of a last name in al-Zigetvari's case might indicate that he was a convert. In Islam and Judaism, converts usually take the name of the founding father of monotheism, Ibrāhīm, or its Hebrew equivalent, Avraham. He also may have been named after a patron, for which there are several possible candidates. During the late sixteenth and early seventeenth centuries, two grand viziers and one sultan were called Ibrāhīm. Also, Abraham was considered in Sufi cosmology to be a microcosmic manifestation of the heart of the cosmos and the receiver of divine knowledge of astronomy.[23]

The name "Tezkireci" was given to those who occupied the position of "certificate maker" within the Ottoman bureaucracy, the official who, under the supervision of the *ra'īsu l-kuttab*, was responsible for the drafting of certificates or deeds. The *ra'īsu l-kuttab*, or chancery of the Ottoman imperial council, was for some time occupied with diplomatic correspondence.[24] It is possible that al-Zigetvari was recruited by the court through the *devshirme*, the Ottoman institution for capturing and training Christian boys, mainly from the Balkans, for military and civil service.

The "al-Zigetvari" portion of al-Zigetvari's name clearly refers us to Eastern Europe. During the mid sixteenth century, Francis I urged his Ottoman allies to attack the Habsburgs; Süleyman the Magnificent, who promoted himself messianically as al-Mujjadid, the renewer of religion, laid siege to the Hungarian frontier fortress of Szigetvar.[25] A week after a lunar eclipse in 1566, as Tycho almost predicted, Süleyman died, and two days later the fortress was conquered and became the Ottomans' administrative center in the Hungarian province. Several decades later, an Ottoman traveler named Evliya Çelebi spent three years exploring Hungary and reported that the new Ottoman rulers had built twenty-five mosques, forty-seven *masjids* (small mosques), twelve *madrasahs* (Islamic colleges), sixteen schools, and nine *khans* (inns for merchants) in the area of Pecs, adjacent to Szigetvar.[26]

Al-Zigetvari's background seems to have been connected to the new Hungarian province. Given that only Christians attended the Latin schools set up under the Ottoman Empire, it is reasonable to assume that al-Zigetvari learned Latin during his childhood at a Szigetvar monastery; only later did the Ottomans capture and recruit him to use his linguistic skills in the court.[27] Al-Zigetvari, then, may have been a captive Hungarian boy incorporated into an Ottoman bureaucracy that utilized his linguistic skills in Latin.

We must also look at al-Zigetvari's choice of title for his translation of Duret's book. Titles, especially in scientific works, are often the threads that connect technical work to the cultural context in which writers, editors, and printers worked. Often, too, they contain clues about the implicit, expected readership. At first glance, the title that al-Zigetvari chose seems incompatible with Duret's book. Duret's title was *Nouvelle théorie*, but al-Zigetvari called his translation *The Mirror of the Celestial Sphere and the Purpose of Apperception* (*Sajanjal al-aflāk fi ghāyat al-idrāk*).[28] Why did al-Zigetvari refuse to use a literal translation of Duret's title? Perhaps it was an innocent attempt to use a mixture of rhymed prose and verse (*saj'*) or an attempt to detach Duret's book from its French context and embed it in a Muslim one (figure 28).

Echoing Titles

Titles often echo, or enter into intertextual relations with, myths or other sources. The word *sajanjal* (سجنجل, "mirror") does not have an ancient Arabic etymology

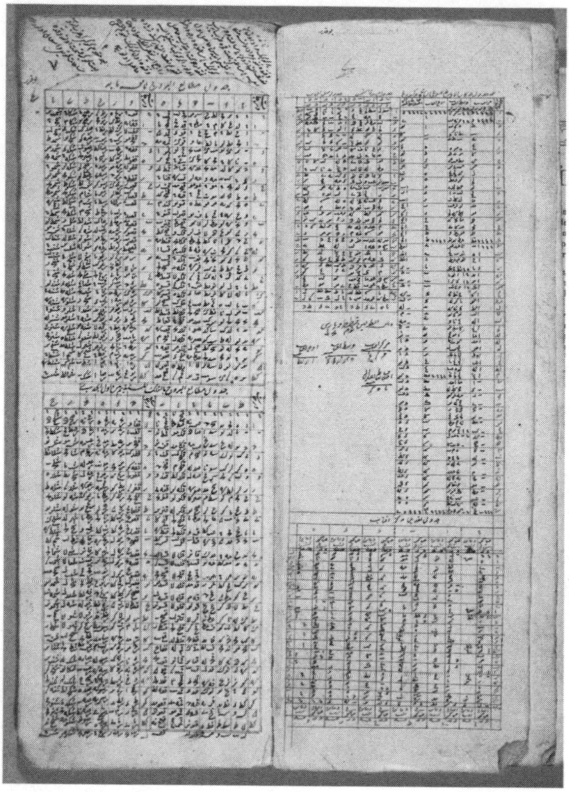

Figure 28. The astronomical tables translated by al-Zigetvari allowed Muslim astronomers, for the first time, to use European astronomical tables, which included astronomical data collected at European vantage points. From Ibrāhīm Efendi al-Zigetvari Tezkireci, *Sajanjal al-aflāk fi ghāyat al-idrāk* (1664), Kandilli Rasthanesi, MS 403, pp. 9a–9b. Courtesy of Ekmeleddin İhsanoğlu.

and was rarely used in the Arab world. Lexicons claim an origin in Byzantine-Greek, thus bolstering the conclusion that al-Zigetvari was not a native speaker of Arabic and that he learned the language in a *madrasah* in Eastern Europe or Anatolia, where the Byzantine culture's influence would still have been strong. Mirrors played a central role in the social and cultural life of the Ottoman city, as objects of art and personal use. There were decorative mirrors, round mirrors with many faces, and mirrors held by Sufi beggars or begging dervishes so that, for a small coin, a gentleman could check his appearance.[29] Mirrors most often alluded to mystic poetry.

Mirrors were symbolic objects in Sufi epistemology. People's interactions with them could construct not just selfhood but also perceptions of a certain melding with God. Any familiarity with this mysticism leads to an encounter with the central conception of the "Hidden Treasure" (or attributes of God) that can only be known through reflection in the human mind: Adam was the first, and perfect, such mirror. Thus the pristine, angelic perception could give access to the "Divine Attributes," which are the hidden things, the secrets concealed in human minds, just as the polished mirror is hidden in a container of felt. In the implicit epistemology, these "Divine Attributes" are hidden in nature, only to be reflected by human acts of mystically acquired apperception. Thus, "Nature," beyond the observable phenomena, can be seen as a subjective construction of the human mind.

In *Mesnevī*, by the great Sufi poet Jalāl al-dīn Rūmī (d. 1231), we read that "since the [mirror's] image of the eight heavens shone forth, the tablet of their hearts had opened to it. One hundred impressions of Highest Heaven, the Sphere of Stars, the Void . . . But what impression this? No! the very sight of God."[30] Rūmī's *Mesnevī* was such a powerful source of inspiration that much of the mysticism that permeates Ottoman poetry makes reference to it in some way.[31]

The rest of al-Zigetvari's title for the Duret translation is familiar-sounding and suggests an acquaintance with Muslim astronomy and the philosophy of illumination. The title echoes those of well-known, older writings on the sciences of the heavens. For example, the astronomer and illuminist philosopher Quṭb al-Dīn al-Shīrāzī wrote *Nihāyat al-idrāk fi dirāyāt al-aflāk* (The Ultimate Perception of the Knowledge of the Heavenly Bodies). Shīrāzī and others exposed their teacher, Naṣīr al-Dīn al-Ṭūsī, to illuminist philosophy, and the latter accordingly renamed a summary of his main astronomical work *Zubdat al-idrāk fi hayāt al-aflāk* (The Essence of Perception of the Form [configuration, behavior] of the Celestial Sphere). Al-Zigetvari may have known such books, the titles of which could have inspired him, because so many Muslim astronomers of this time read these works. But if we look into the deeper context of parallel wordings and intertextuality, we find that al-Zigetvari participated in a contemporary trend of natural philosophy.

Seeing beyond the Natural Phenomena

Late in the twelfth century, Shihāb al-dīn Suhrawardī (d. 1191), a court philosopher of the Ayyubid dynasty, launched his devastating criticism of Peripatetic philosophy. His illuminist philosophy was not a kind of theosophy or mystical philosophy, as some once suggested,[32] but emanated from rigorous logic.[33] Mentioning his intellectual sources, Plato and Hermes, Suhrawardī argued that the existence of a valid mental distinction did not imply the corresponding existence of an actual distinction in concrete things. In other words, man constructed such metaphysical entities as "essence," "necessity," and "causality" as "beings of reason" (*i'tibārāt 'aqlīya*, اعتبارات عقلية). Instead of relying on "beings of reason," Suhrawardī looked for an intuitive perception of nature. As the angelic, pristine perception in hermeticism and Sufism, intuitive perception would arise as a spontaneous sensation and epiphany, before the working of the faculties of reason and senses that perceive new objects. Astronomy was thought to be a suitable place to show that the combination of reason and senses was an inadequate explanation of nature.

Suhrawardī's work conveys not mystical experience but a rational structure that records rational proofs found after the mystical experience. In Suhrawardī's analogy, rational proof is to mystical experience as astronomy is to observation.[34] "Just as by beholding sensible things we attain certain knowledge about some of their states and are thereby able to construct valid sciences like astronomy, likewise we observe certain spiritual things and subsequently base divine sciences upon them."[35]

Perceiving the world through deep immersion in the mystical experience was, for Suhrawardī, like opening a window to a place that reason and senses cannot reach. Suhrawardī's emphasis on "illumination," "radiance," and "light," coming from hermetic-Pythagorean learning, gave a central place to the sun. It became the source of the corporeal light through which we can trace hidden, incorporeal light, and the eternal, circular movements of the spheres are the intermediary through which the timeless and unchanging complexity of the incorporeal light is expressed.

Suhrawardī lays down two premises: on the one hand is the inadequacy of the weak constructions of the mind that are made by reason and senses and by deductions from experience; on the other hand is the incorporeal light that can be reached by various tools of natural philosophy—specifically, *idrāk* (ادراك). The latter word can mean "perception," yet for Suhrawardī it had a more complex philosophical meaning. His aim was to construct a unified epistemic object that would supplement what reason and senses could not reach. The notion had already been used by Avicenna,[36] and later by others, to describe mainly a sensual perception, but Suhrawardī took it one step further and elevated it to the status of an epistemic object that could judge ultimate truth. *Idrāk* is an intuitive mode of cognition, a direct knowledge of something, whether through sensation or intuition. Through apperception, we not only can see the visible movements of the stars, but also can predict events, because the forms of events are inscribed in the spheres and are accessible to human beings under special conditions and with special tools. In *The Book of Radiance*, Suhrawardī writes: "they (the celestial spheres) possess universal knowledge, know universal laws, and thus know every form and effect that is in this world. When a celestial sphere arrives at a specific point [in its orbit] it becomes cognizant of whatever comes about as a result of achieving that point. In short it comprehends all temporal causes of effects from past to future, as well as the present."[37]

Suhrawardī's work constituted a new natural philosophy in which *idrāk* unifies the practices of acquiring knowledge across mysticism, astrology, and astronomy. But Suhrawardī was not the only one who was concerned with such matters. Whereas he approached the problems from an illuminist point of view, Ibn al-Haytham (Alhazen; d. after 1041), a near-contemporary natural philosopher, was concerned with optics and the limits of any rational understanding of nature. Like Suhrawardī, he advocated sensual-intuitive perception and intensively employed the object of *idrāk*. For example, in his work in optics, the object served to indicate something that lay behind our perception of the physical world and was not confined to the internal features of optical processes or the organ of sight. *Idrāk*, to Ibn al-Haytham, was pure sensation.[38] He was also one of the first critics of the Ptolemaic model and especially its failure to reconcile the physical world with the planetary model, as in the Ptolemaic equant's violation of the geometric rule that uniform revolution is measured only from the centers of spheres. Ibn al-Haytham argued that the source of this inconsistency was Ptolemy's methodology, which led Ptolemy to depart from his own theory. And because Ptolemy knew this, his excuse on the grounds that the inconsistency does not affect the observed movements of the planets would be acceptable if one believed there is a "true" nature of the heavens (*kunh haqīqatihā*),[39] over and above Ptolemy's construction of the heavens. For Ibn al-Haytham, *idrāk* in optics and astronomy was the same: it would facilitate our perception of the "true" nature that could not be perceived by reason or senses.

The most important followers of Suhrawardī and Ibn al-Haytham were al-Ṭūsī and Quṭb al-Dīn Shīrāzī, members of the Marāgha Observatory.[40] Shīrāzī's commentary on Suhrawardī's writings became the vehicle through which the philosophy of illumination was later understood. For the Marāgha School, the main problem was the physical and mathematical disharmonies in the Ptolemaic system, and Shīrāzī's investigations provided a non-Ptolemaic model for uniform circular movement. The breakthrough accomplished by the astronomical inquiries of al-Ṭūsī and Shīrāzī was to use Suhrawardī's *idrāk* as an episteme beyond reason and senses, one that would unify observed data with their physical and mathematical models. *Idrāk* would fill the gap between what is observed in the heavens and the limited ways of describing it. The attempt to fit observation to circular motion, in fact, would be partially achieved later, in fourteenth-century astronomy, by Ibn-Shāṭir, who got rid of the equant in his model for the moon.

Idrāk did not remain just an episteme in al-Zigetvari's scholarly world of astronomers and natural philosophers. It also dealt with the practice of acquiring perfect, unified knowledge in relation to rituals of self-development. Rūmī and other Sufi poets and philosophers, including Ibn Al-ʿArabī, claimed there was a need for obtaining and cultivating a pure consciousness that could get to so-called *wajdat al-wujūd*, the unity of existence. In this endeavor, *idrāk* became a means and illumination became the purpose. *Idrāk* was an epistemic object that could unify the believer with nature and subsequently with God. It therefore entailed ritualistic practices: *dhikr* (recollecting, a repetition of divine names or religious formulae), *ta'ammul* (meditation), *tafakkur* (pondering), and *tawajjuh* (strong concentration of master and disciple on each other). Sufi practices were formed by the striving for intuitive perception. The practitioner would elevate his spiritual experience and gradually "clean" the mirror of his heart to a level at which God "lifts the curtain" and "unveils" the heavens. At this point, the practitioner would reach the highest spiritual experience, in which God and the heavens were completely reflected in the mirror of his heart.[41] Rūmī's *Mesnevī* had already been a lyric command in this regard: "Show me things as they really are. Man is a mighty volume; within him all things are written, but veils and darknesses do not allow him to read that knowledge within himself."[42]

To simplify, such things as *idrāk*, illumination, mirrors, astrology, radiance, and hermeticism make up "the Muslim emblematic world." Each constitutes, in the mirror of the heart, a reflection of the "true" heavens. *Idrāk* made an impact in areas of knowledge associated with astronomy. Taqī al-Dīn (d. 1585), chief astronomer of the Ottoman Observatory, was also well-known as a mechanic specializing in clocks for a variety of uses (as described in chapter one). In the introduction to his treatise on astronomical clocks and their application to observation, he claimed, using Hermes as a source of inspiration, that the conditions for calling forth the pure truth about the motion of the heavenly bodies is *idrāk*.[43] He placed a condition on those who would try to replicate the astronomical instrument: they needed to have the power of *idrāk* to be able to set and measure the motions.[44]

Sufism, with its strong social position, carried forward and perpetuated the ideals of *idrāk*, illuminationism, and the mirror motif in seventeenth-century Ottoman art and architecture. Sufi religious orders (*tarīqa*) imposed hierarchical structures on social groups, from sheikhs to dervishes and their sympathizers. Ottoman Sultans seeking hegemony over Anatolia relied on the Sufi orders for support in consolidating newly acquired territory. In return, the Sufi orders received gifts of land and endowments for lodges (*zāwiya*), in which members of the orders practiced rituals and taught the unity of existence through apperception. The lodges, built around centralized, harmonious spaces, created harmonious atmospheres where practitioners could circle and meditate, in a "polishing of the heart." There, from the sixteenth century onward, Sufis and students practiced a special style of calligraphy that used mirror-reflection motifs to express holistic perception (figure 29).[45] We also find a strong Sufi trend in areas besides popular culture. The loftiest government circles and centers of astronomical activity were infused with it.[46] Later in this chapter, we see that al-Zigetvari presented his translation of Duret's book to the *munajjim bāshī*, or court astrologer-astronomer. The man who held that position beginning in 1667, three years after publication of al-Zigetvari's *Sajanjal*, was Ahmet Munajjim Bāshī, born in Salonika and educated in traditional scholarly subjects, including *fiqh* (Islamic law) and the sciences, mainly astronomy and astrology. But he was also well versed in Sufism, such that although he carried the title "chief astronomer," he was mainly known to his contemporaries as an outstanding Sufi poet.[47]

Thus, not just natural philosophers and astronomers but mystics, Sufi dervishes, and poets also concerned themselves with illumination through apperception. And this was not merely a structural connection among various concepts—first, among various medieval sources, and then between these sources and our protagonist, al-Zigetvari. In fact, there was a personal connection within a "cluster of intellectuals."[48] Al-Shīrāzī's commentary was the main vehicle of Suhrawardī's thought; al-Ṭūsī was familiar with Suhrawardī through his own student, Shīrāzī, and through other sources;[49] and Shīrāzī was personally acquainted with Rūmī.[50] Words and myths used in al-Zigetvari's title for *Sajanjal* reflected the concerns of these astronomers and philosophers. Al-Ṭūsī, al-Shīrāzī, and al-Zigetvari were inspired by and acted within a contemporary theory of knowledge. It was no coincidence that they all used the term *idrāk* in the titles of writings. Alternative terms were available—for example, *arṣād* (ارصاد), meaning "observations," a basic part of their practice of astronomy. But in their works, *idrāk* (as in "the essence of *idrāk*," "the ends of [ultimate] *idrāk*," and al-Zigetvari's "the purpose of *idrāk*") described the more fundamental philosophical action of apperception. In their scientific practices, these men were engaged both in observations and mathematical models and in mystical practices aimed at refining their consciousness. More than merely solving mathematical and physical inconsistencies in the

Figure 29. Art and *idrāk*. In the *muthannā*, or "double" style of calligraphy (left), each half of the design is a mirror image of the other. The *basmallah* in the *thalūth* script (right) is written in the shape of an ostrich. The calligraphy shown in this illustration is not really a type of script in itself, but consists of a text in one of the standard scripts, such as *naskhī*, worked into a pattern in which one half is a mirror image of the other, or is pictorial calligraphy in which the text (usually a profession of faith, a verse from the Qur'ān, or some other phrase with religious significance) is written in the shape of a bird, animal, tree, boat, or other object. The act of reading usually involves identifying words (a sensory and cognitive act) and grasping the meaning (an act of reason) that is delivered in the words. The holistic intuitive perception, or *idrāk*, allows an appreciation not only of the details of the letters and words but also, more importantly, of the constructed holistic image of the bird.

Ptolemaic system, they sought a new harmony of the scholar and his cosmos—a microcosmic, human mirroring of God's universe in the heart of the practitioner, achieved by using pristine angelic and hermetic apperception.

The theory of knowledge also entailed practice, and theory and practice also had implications not just for epistemology but for intellectual culture as well. It is worthwhile considering other sorts of links between fellow scholars and officials both at court and in high-profile observatories. Not only did the men discussed above seek to reconcile epistemological inconsistencies arising from their passions about the act of perceiving nature, but the striving for *idrāk* itself became a sign of sincerity and an indication of trustworthiness.

From evidence farther east in the Muslim world, we see that an unintended consequence of all this was the forging of bonds of trust among working colleagues in similar scholarly contexts. In a letter written by Ghiyāth al-Dīn Jamshīd al-Kāshī, a mathematician and member of the fifteenth-century Samarqand Observatory, he describes the politics, culture, and practices that formed the backdrop of the astronomical texts produced in Samarqand. He laments the lack of wholeness in the learning and skills associated with theoretical and practical astronomy. For him, the division of labor between model builders, calculators, observers, instrument builders, astrologers, and mathematicians was not for the sake of efficiency but a reflection of fragmented knowledge.[51] As a result, mutual scientific trust was not strong. But one person who gained his trust was a Sufi practitioner in the observatory. The Sufi was a pure seeker of truth who, in seeking unified knowledge of a unified nature, erased the boundaries between disciplines and the differences in professional prestige. Al-Kāshī perceived him as striving to advance knowledge rather than his own social status, and therefore as more trustworthy than others. Through this fascinating glimpse we can see that a Sufi's sincere passion for harmonious knowledge, through *idrāk*, concerned both epistemology and trust.

In the same vein, we encounter the quintessential Ottoman intellectual of the seventeenth century: Kātib (Kâtip) Çelebi, an explicit follower of Suhrawardī.[52] Besides his work in cataloguing, Çelebi also undertook to reconcile the conflicts between theology and philosophy, and between rationalism and mysticism, by following Suhrawardī's theory of knowledge: to reach the heights of study and investigation, scholars should not "confine themselves to one branch of knowledge."[53] Moreover, Çelebi took a further step and set forth principles for an appropriate intellectual culture, suggesting that the "sincere" striving for unified knowledge not only should be a matter of the knower and the known, but should be applied to the practices of trust between scholars. For him, someone who sincerely looked for knowledge would be less argumentative and divided in his thinking, and he provided advice on how to practice trustworthy scientific activities.[54] Thus, the mystical trend of knowledge for the sake of knowledge became a pillar of trust among scientists. Practices for achieving the ultimate standard of truth, such as *idrāk*, became an expression of sincerity and an object of trust.

Al-Zigetvari's Translation

Having established some of the cultural and intellectual sources in Sufi cosmological poetry and illuminist natural philosophy, we can make some guesses as to why al-Zigetvari chose to translate Duret's book. An analysis of al-Zigetvari's translation takes us beyond the not too useful dichotomies occult/scientific, astrology/astronomy, and mysticism/logic. We are instead considering a text that meant to overcome logical and scientific aporia. Al-Zigetvari read Duret's book within a context of particular cultural values, validating the text by presenting it through *idrāk*.

Sajanjal al-aflāk fi ghāyat al-idrāk is divided into two parts: the translation proper and, more important to us, al-Zigetvari's introduction. The content of Duret's book was predominantly astrological, and the Copernican system was relegated to an aside. Al-Zigetvari's introduction makes little mention of astrology (apart from an astrological diagram) and, instead, amounts to his own reading of the history of astronomy.

Al-Zigetvari states that he found Duret's book and independently elected to translate it into Arabic: that is, only after completing the translation did he show it to the *munajjim bāshī* (the sultan's chief astronomer-astrologer), Mehemet Çelebi.[55] At first, Mehemet Çelebi did not approve of it, stating that "Europeans have many vanities similar to this one." But when al-Zigetvari prepared an almanac based on Duret's tables, the *munajjim bāshī* saw that it conformed to the authoritative tables prepared by the Muslim astronomer Ulugh Beg, and thus he became convinced of the value of the French work. Al-Zigetvari had a copy made for himself and made a gift of another copy to Çelebi. The lengths to which al-Zigetvari went to convince Mehemet Çelebi of the value of Duret's book suggests that the court astronomer was firm in his Ottoman sense of superiority, yet was also appreciative of sheer utility.

Harmonizing the Worlds

Al-Zigetvari's descriptive and encompassing history, as laid out in the introduction to *Sajanjal*, reflected the quest for a harmonious description of nature through *idrāk*. The introduction traces astronomy from the Greeks through the Muslims to Copernicus. The translator notes the methodological shortcomings of ancient astronomy based on observation: the improvements made by the Muslims came about "by means of the *zīj*,"[56] that is, the *azyāj* school of computational astronomy. This was the first phase of testing Ptolemaic data by conceiving, organizing, and executing new series of observations (*arṣād*). The second phase, which involved questions of planetary configuration, was called *hay'a*. The emphasis of the *hay'a* school's newly conceived problem-solving program was primarily theoretical rather than empirical; the program consisted of seeking, or urging others to seek, reconciliation between the Ptolemaic "mathematical" hypotheses assumed to be supported by observational tests and the theories of cosmology, physics, and natural philosophy. A. I. Sabra suggests that this school perceived the existence of "true" (*ṣaḥīḥ*) configurations that were in harmony with the accepted principles of physics.[57] Al-Zigetvari claimed that the source of progress in Muslim astronomy was its quest for a more harmonized system that would reconcile the physical with the mathematical.

Al-Zigetvari's history of astronomy in the introduction to *Sajanjal* stressed that all astronomers sought out harmony and searched for a way of unifying perception that would explain technical inconsistencies. We should note that, in this narrative, as Arabs took over the mantle of astronomy from the Greeks, no mention is made of different religious and ethnic backgrounds. Ekmeleddin İhsanoğlu's translation of the introduction shows that, for Al-Zigetvari, the astronomers existed simply as cosmopolitan individuals and not as representatives of cultures.

> In 1461 Peurbach and Regiomontanus found mistakes in Alfonso's *zīj* (astronomical tables). Although Regiomontanus had initiated his observations in order to correct the tables, he did not live long enough to finish his work. A few years later Nicolaus Copernicus, who was very successful and a superior scholar, discovered the mistakes in Alfonso's tables and, realizing that the foundations for the tables were unsound, formulated a new solution in 1525. He did so by relying on the observations astronomers had made throughout the ages. The defect in question was the following: the eighth sphere moves with the ninth sphere by an oscillatory motion *(raqqāsiyya)* of 40,900 years. This motion is equal to two small circles moving from West to East in the depth of the ninth sphere. This situation is against the observation of the majority. Later, Copernicus laid a new foundation and compiled small tables supposing that the Earth is in motion. These tables were used thereafter until Tycho Brahe made observations with numerous excellent instruments and began to correct the tables of Copernicus. Relying on his observations, he also wrote the tables of the Sun, the Moon, and the fixed stars . . . Afterwards, the scholar Kepler compiled tables for all stars based on Tycho's observations, and named it the Rudolfine Tables . . . The eclipses of the Sun and the Moon did not conform to this tables . . . later I, Ibrāhīm Sgizetvari, known as Tezkireci, translated the tables which the scholar Duret compiled after thirty years of observation based on the tables of Lansbergen. The tables were based on observation according to

> the longitude of Paris. It was compiled by benefiting from the old Julian and the new Gregorian calendar as well, accepting the Earth as motionless.[58]

Here, the history of natural philosophy is a chain of knowledge transmitted from one scholar to another, from one generation to another, smoothly and continuously—a search for harmony in a mutually embraced world history. No revolution is mentioned in either period; it was merely an evolution of knowledge.

Al-Zigetvari does seem to have been aware of the disagreements between the heliocentric system of Copernicus and the systems of the pre-Copernicans, and between Copernicus and Tycho, and his annotated diagram presents such differences. Yet, he did not take a clear stance on which system he found more convincing. Duret supported the argument that the earth is motionless, and al-Zigetvari seemed to care more about the harmony of the heavens and less about the location of the earth and sun in the cosmological structure.

Of course, al-Zigetvari was not reading Copernicus directly but was getting it through the mediation of Duret. Both Duret and al-Zigetvari refer to a writing of Copernicus from 1525, which is closer in time to the *Commentariolus* than to the *De revolutionibus*, works that reflect the changes in Copernicus's motives. The discussions about his motives in our own time might offer some useful insights here. Some scholars, such as Thomas Kuhn, relying on *De revolutionibus*, think that Copernicus worked to eliminate mathematical contradictions and improve overall harmony through geometric aesthetics.[59] Others, such as Noel Swerdlow, relying on the *Commentariolus*, think that the astronomer sought answers to physical and philosophical problems surrounding the celestial spheres and that he had intellectual roots in the Marāgha School, and especially al-Ṭūsī and Ibn-Shāṭir.[60] One might first assume that al-Zigetvari did not follow predictably in the footsteps of Marāgha. He chose instead to present the problem differently, as questions about the eighth and ninth spheres, as seen in the extract from the introduction given above. He mentions neither the heliocentric nor the equantless features of Copernicus's model but focuses on the question of values for the movement of trepidation and precession.[61] In considering his take on Copernicus, we cannot expect an intellectual model or a historical motive, as we interpret such things today. The quest for the harmony of the heavens by the Marāgha School modeled al-Zigetvari's interests in the geometric harmonization of trepidation and precession.

The matter of the spheres represents al-Zigetvari's reading of Duret, who in turn also read and extracted ideas (indirectly) not necessarily just from Copernicus's writing but from secondary sources of the late sixteenth century. As mentioned earlier, Duret's book emphasizes the role of post-Copernican astronomers such as Tycho, Lansbergen, and especially Kepler, and omits Copernicus, ostensibly for censorship reasons. Robert Westman deals with these secondary sources for Copernicus and has tried to do away with any supposed division between Copernicans and anti-Copernicans in the period when the Copernican system was first received.[62] He argues that the process at that time, especially in Wittenberg circles, was elastic: astronomers such as Reinhold and Peucer did not accept the premise that the earth is in motion, but they did accept and use Copernican mathematical devices. In contrast, astronomers such as Rheticus accepted the moving-earth premise, not for the mathematical devices, but for the purpose of reconciling inner conflicts and bringing together a harmonious universal whole.

The same kind of elasticity exists in the case of al-Zigetvari (figure 30). He situated the problems in astronomy in the technical arena of Ptolemaic discrepancies, as did many of his predecessors, but he mainly presented the motives of Copernicus as a problem of aesthetics, just as Kuhn suggests. The reason for his doing so may very well be related to the way al-Zigetvari received his Copernicus. Based on the extant text of Duret, we can suggest that in reading Duret, al-Zigetvari was imbibing the latter's understanding of Copernicus via Rheticus's *Narratio prima*. Moreover, the entire point of this narrative up to now has been to show how very broad was the appeal to al-Zigetvari of the aesthetic explanations regarding harmony that we find in holistic, illuminist Sufism and the type of Marāgha culture in which it was embedded. Al-Zigetvari's epistemological elasticity and his aesthetic take on post-Copernican astronomy can also be thought of in the manner suggested by Westman in his discussion of Rheticus—that is, as a way to reconcile the inner conflicts of a Hungarian convert in the service of the Ottomans.

The Utility and Appeal of al-Zigetvari's Work

Al-Zigetvari tells us about the practical accomplishments in translating Duret.

> I worked out all the mean positions of [Duret's] tables again according to the evidence and rearranged its shape on the same longitude. I expounded it in Arabic. This ephemeris [his own development] was sexagesimal and its mean positions were universal; I arranged it according to the constellations of the zodiac. I turned the tables into an ephemeris hitherto unseen in shape and brevity, entitled it *Sajanjal al-aflāk fi ghāyat al-idrāk* . . . I also compiled a work

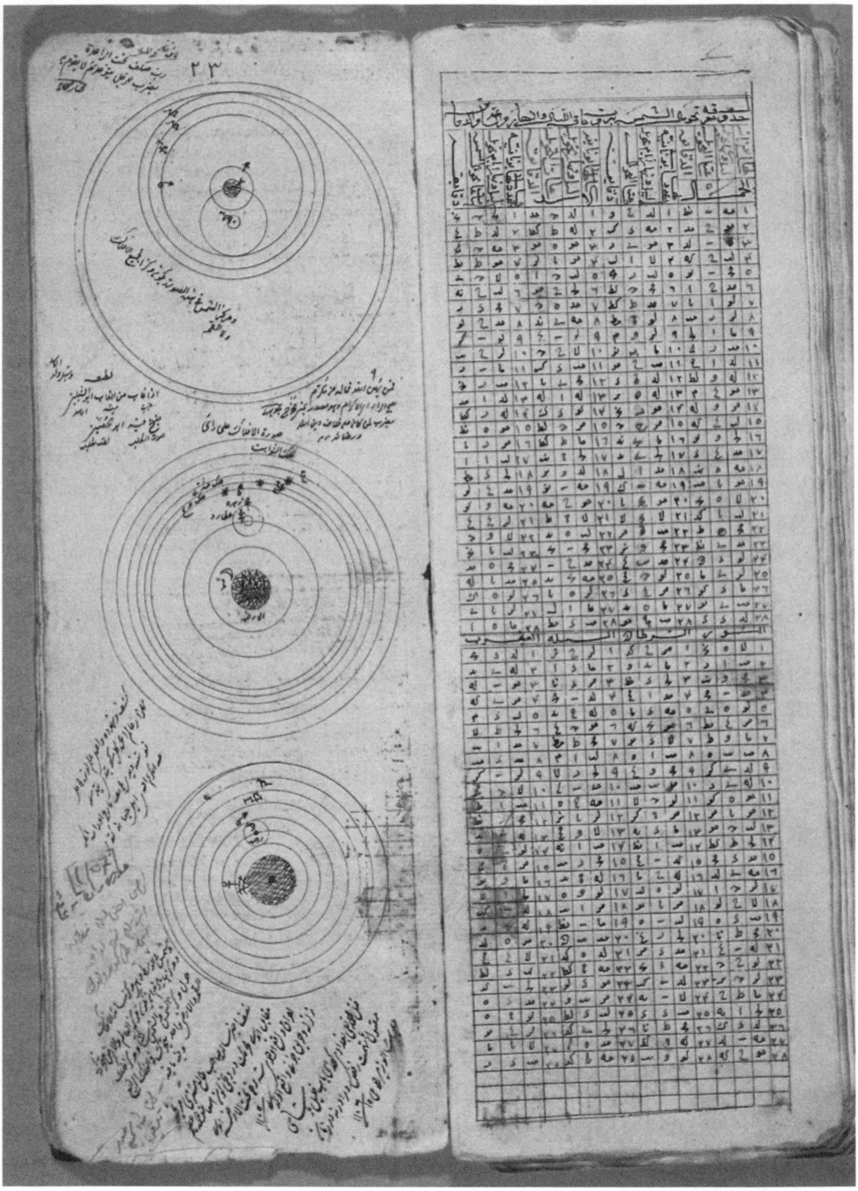

Figure 30. Al-Zigetvari's diagram of the world-system (left), shown with its companion astronomical tables (right). From Ibrāhīm Efendi al-Zigetvari Tezkerci, *Sajanjal al-aflāk fi ghāyat al-idrāk* (1664), Kandilli Rasthanesi, MS 403, pp. 22b–23a. Courtesy of Ekmeleddin İhsanoğlu.

concerning the universal calendar, which came out to be more graceful and succinct than all calendars.[63]

Because the growing number of astronomical tables created divergent and incompatible figures for the positions of the heavenly bodies and the latitudes of cities, al-Zigetvari compared the numbers in various tables, Muslim and European, and calculated their averages as the accepted numbers. Instead of an isolated, or singular, use of specific observational data, al-Zigetvari's treatment suggests a holistic and conforming perception (or apperception, *idrāk)* of the locations of the heavenly bodies. Al-Zigetvari's adjustment of the imported *Nouvelle théorie* to the Ottoman practice of astronomy followed the astronomical genre of the Muslim world known as *azyāj*, a record of observations whose purpose was specifically to update the *zīj*. Over centuries of updating, it was evident that the Ptolemaic values of precession and procession—solar apogee, solar equation, inclination of the ecliptic, and all Ptolemy's mean motions—were no longer acceptable.

But most important for the discussion here is that al-Zigetvari's universal and holistic treatment reflected his illuminist and mystical background: he avoided certainty and any clear notion of causality by looking for a universal and unified understanding of nature. Moreover, the arrangement of the astronomical tables according to the latitudes and longitudes of various European locales presented a unified spatial perception that went beyond the parochial spatial perception in medieval Islamic tables.

Yet, in the same way that the heliocentric system was at first rejected and disregarded in Europe, the Ottoman scientific authority disdained al-Zigetvari's presentation of Duret's post-Copernican astronomy. As we have seen, initially the *munajjim bāshī* rejected *Sajanjal* as "European vanities"; only after al-Zigetvari convinced him that the book agreed with the prevailing astronomical tables of the Muslim world did the chief astronomer give his blessing to the translation, saying, "You have erased my suspicions; now I have full confidence in the tables prepared by Europeans."

Why did the chief astronomer initially resist the translation of new research? After all, was this sort of knowledge-gathering not part of his job? The resistance was related to the utilitarian ends to which Duret's book could be applied. The scientific environment surrounding al-Zigetvari was not interested just in theoretical astronomy. In fact, astronomy and astrology both possessed a utilitarian character in the Ottoman context. Astronomers of local mosques used astronomical tables for timekeeping, and astrologers used them to draw up horoscopes.

The illuminist philosophy and mysticism in al-Zigetvari's intellectual makeup inspired him to pick Duret's book from the court library and to translate it, but it was its astrological utility that won over the *munajjim bāshī*.

Al-Zigetvari's translation marked the first clear mention of the new astronomy in general and the Copernican system in particular, but he mentioned the Copernican system only along with other world-systems and was more interested, for utilitarian reasons, in the astronomical tables than in the structure of the universe. It took some centuries for a full acceptance of heliocentrism. A translation of Edmund Halley's (1656–1742) short Latin work *Astronomiae cometicae synopsis* (1705) appeared in Istanbul under the title *Tarjama Muqaddimāt Wâhiya fi Ahwāl al-Nujūm zū Zuāba*. The manuscript mostly dealt with the question of comets, but the anonymous translator accentuated the unambiguous truthfulness of heliocentrism by completely omitting any mention of the other systems and by presenting, on the last page, a clear illustration of the heliocentric cosmology.[64] Not until the nineteenth century, then, did Eastern Mediterranean scientific culture completely abandon the Ptolemaic and Tychonic cosmological models and fully accept the Copernican system.

In 1635, Richelieu's "cosmographer," Noël Duret, published *Nouvelle théorie*, a book that presented convenient astronomical tables based on the works of various astronomers, including Copernicus. Sometime between 1637 and 1642, thanks to Richelieu's fondness for art and scholarship, the book was delivered by Harlay de Cèsy to Constantinople in exchange for Arabic and Persian manuscripts sent to Paris. The *Nouvelle théorie* lay on the shelves of the court library for twenty years, until a Latin-educated Hungarian convert to Islam, Ibrāhīm Efendi al-Zigetvari Tezkireci, came across it. Inspired by Sufism, illuminist philosophy, and astrology, al-Zigetvari showed the *munajjim bāshī* his translation of Duret's work. The *munajjim bāshī* gave his approval only when he was convinced of the astrological utility of the astronomical maps. Al-Zigetvari's work of translating and providing a commentary on Duret's Latin book played against the backdrop of *idrāk*. Through a mystical episteme, al-Zigetvari interpreted his readings of the problems of astronomy.

This narrative about a Parisian publication and its translator in Constantinople raises questions about the prevailing historiographic view of the period. Many scholars have pointed out that Islamic natural philosophy was an important bridge to the rise of the new natural philosophy in Europe. However, little has been written on the philosophical and scientific practices in the Islamic world in the early modern period. The neglect is not surprising.[65] Historians of early modern Islamic natural philosophy have asserted a sort of disequilibrium—a "progressive" Europe and a "backward" Islam—blaming the supposed lethargy of Islamic natural philosophy on the defeat of rationalism and the rise of Islamic mysticism, or Sufism.[66] Some have linked the "decline" to two twelfth-century philosophers: al-Ghazzālī and Suhrawardī, who wrote devastating critiques of the notions of rationalism, causality, and certainty.[67] Al-Ghazzālī (d. 1111) took a theological angle in his *The Incoherence of Philosophers (Tahāfut al- falāsifah)*, which tried to demonstrate that philosophers are unable to prove religious truths and have no tools to understand God's reasoning. The second attack, by the illuminist Suhrawardī, took a logical angle. His *Philosophy of Illumination (Ḥikmat al-ishrāq*; literally, the *Oriental Philosophy*) suggested that causality and other Peripatetic categories were only a construction of the human mind. This line on Islamic natural philosophy also asserted that the works of al-Ghazzālī and Suhrawardī gave rise to and played a central role in Sufism, which thus sent Islamic natural philosophy into a sort of deep sleep, until eventually confronting the new European natural philosophy in the age of Western colonialism. It is to this sort of master narrative that recent scholars have responded.

The widely held notion of a decline in Islamic natural philosophy due to Ghazzalian and illuminationist philosophies does not fit the world of al-Zigetvari and Duret. Actually, what we see is that Sufism produced epistemic objects and practices that aimed to solve inconsistencies in astronomy. It inspired an epistemological search for a harmonious and unitary nature and, more practically, it proposed rituals and practices to gain a refined consciousness that would allow a special apperception (*idrāk)*. Finally, Sufism motivated a search among astronomers for a way to trust each other and their projects and to foster, in the context of their religious passion, "true lovers of truth" who looked for knowledge for the sake of knowledge.

Ever since Frances Yates's account of the hermeticism of Bruno, European historiography of natural philosophy has gone through major revisions regarding the role of the occult in the rise of early modern natural philosophy. It makes sense to include in this revisionist program the kind of intellectual framework provided by al-Zigetvari, Sufism, and *idrāk*. As we have seen, Sufism was not an impediment to scientific activity, but a driving force for further exploration.

With Yates in mind, one can suggest that al-Zigetvari's translation of Duret's work represents Ottoman occult illuminist traditions that stimulated an interest in post-Copernican astronomy. It also challenges the negative historiographic approach to Sufism. Now we might ask, How did Sufism and occultism stimulate new scientific work and drive the reading of post-Copernican astronomy? The answer, in part, is that al-Zigetvari used a holistic approach and simply included post-Copernican astronomy in his inclusive, flowing account of a world history of science. He presented the new cosmology not as a break from tradition but as a continuum, which touched on the Marāgha School's effort to harmonize the heavens not just through mathematics and observations but also with intuitive perception.

Muslim astronomers dealt in precision: they updated and reformed Ptolemaic astronomy through new tools of observation, high-quality mechanics, mathematization, new theories of optics, and a persistent discussion of Aristotelian physics. However, these were all part of a larger picture of intellectual activity and part of a larger project that was embedded in the culture. Thus, the history of Muslim astronomy from the Marāgha School onward should be seen not simply as a reaction to the Ptolemaic system, but as a reaction against both untrustworthy scholarship and approaches to knowledge and observation that were not properly harmonized. The scholar needed to refine the mirror of his heart so as to be fit to work in God's universe. Sufism and illumination were not an obstacle and could serve as sources of inspiration for the translation of the first Arabic text to mention Copernicus.

From "Incommensurability of Cultures" to Mutually Embraced Zones

An exclusive depiction of the local development of science has been qualified ever since scholars began exploring the rich interconnections between travel and science. Studies concerning the discovery of the New World and the rise of natural history, however, focused on European travelers as they headed westward, to the New World, and left unnoticed the cross-cultural exchanges that took place in the "Old World," in the Near East. Scholarship of travel and science, ironically, has further contributed to the notion of a great, insurmountable divide between European and Near Eastern intellectual cultures. Yet, the two cultures embraced each other in the Eastern Mediterranean, where practitioners searched for evidence in the long-lost past of ancient Greek, Arabic, Egyptian, Jewish, and Chaldean centers of knowledge and, in passing, circulated and encountered objects that carried the news of the new heliocentric-related cosmologies.[1]

The force behind the propagation of scientific ideas, unexpectedly, did not stem from intellectual networks and agendas, but rather from mundane networks of contact. Pirates violently crossed cultural and geographic boundaries, capturing travelers and exchanging them as goods in various ports. When they discovered scholars among their captives, they marketed them to scholar-masters who needed help in translation and research. Scholar-captives demonstrated their particular proficiencies, giving their masters access to knowledge and texts beyond their own cultural horizons. In exchange, captives gained new scientific knowledge that they could apply if and when they returned to their homes.

Patronage relationships also disseminated scientific knowledge. Men of means sent travelers in search of pristine textual objects or biblical ur-texts. Guided by the perspective of European thinkers who located the authority of truth about cosmologies in nonhuman artifacts, pristine objects, or ur-texts, these travelers crossed cultural boundaries, connecting European and Near Eastern scientific cultures and functioning as axes of cross-cultural exchange. At the same time, they introduced the new discoveries in Europe to people in the Near East.

Traveling Sephardic Jews functioned only slightly differently. They moved among Karaite and Jewish communities, where they collected ancient Hebrew manuscripts and brought them back to Hebrew printing presses in Europe. Acting as "bridges of trust" between the Near Eastern scribal and European print cultures, they helped circulate ancient manuscripts in Europe and, at the same time, got printed books and astronomical knowledge into the hands of members of the Jewish communities in the Near East.

Academic travelers also played their part. Under the guidance of high churchmen or independently, they collected manuscripts and conducted astronomical experiments—measuring latitudes from Ptolemy's hometown and comparing their results with current astronomical tables. Local astronomers, high clerics of the Greek Orthodox Church, and European ambassadors facilitated the search for pristine measurements and, consequently, the resolution of theological and scientific controversies. In passing, they also propagated post-Copernican astronomy across cultures.

Finally, diplomatic networks circulated scientific objects, if in a more circuitous manner. As displays of power, European and Near Eastern courts exchanged gifts, including instruments and books of natural philosophy. Court bureaucrats then gained access to scientific sources of cultures that otherwise stood in opposition to one another. Working within the bureaucracy, they were licensed to translate works and introduce them to local cultures.

A variety of human carriers, then, moved through networks that allowed them to circulate knowledge about nature, science, and philosophy. They were also the agents that put scientific objects—telescopes, compasses, quadrants, mechanical clocks, maps, globes, and books of natural philosophy—into motion, advancing their own specific agendas unsystematically, often before any intellectual need was made known. Allusions to a new cosmology were either a by-product of larger interests or mere happenstance.

Momentary exchanges of foreign objects were connected to enduring cultural conceptions through a larger, mythical framework. Despite the lack of means of communication—different languages, religions, and readings of objects—interactions took place against a set of myths about a single origin for humanity that, in the distant past, had the purest form of knowledge about nature. Myths of ancient theology, in turn, were called on to judge the validity or falsity of heliocentrism and played as tools for objective assessment of the new cosmologies.

What can seem like an inarticulate cultural dialogue among people of different cultures approaching the same objects with different agendas was in fact a meta-cultural dialogue. Everyone engaged in the exchange was acting within a certain historical and cultural framework, arising from hermetic and biblical arguments concerning a single universal origin of all civilizations and undergirding all cultures. For instance, motivated by the challenge of the coming (1590) Islamic millennium, Muslim astronomers used hermeticism to uncover cosmic secrets. At the same time, European travelers noted in their itineraries that they tried to emulate the legendary travels of Pythagoras, in the hope of finding hidden sources of the ancient lore of heliocentrism. On their travels, they also discussed the secrets of the new cosmology with locals, convinced that they could recover hints concealed in biblical ur-texts and in an ancient, lost, universal language, in which the measurements of the universe were given to the ancient Egyptians, to Noah, and to Solomon. Sephardic Jews shared their belief that the search for ancient wisdom was based in ancient cabalistic Hebrew texts. Ottoman bureaucrats translated European works on the new cosmology and presented them to readers as books of mysticism that crystallized the perception of nature through illumination and intuition, an approach that prevailed among Sufis. All these practitioners approached the objects through different cultural angles but within the same framework of *prisca theologia*, a universal theology, which turned out to be a shared meta-language.

Once we expose such cross-cultural networks of exchange, far-reaching historiographic implications come to the fore and underline the limits of cultural history. The contextualization of science within a deep cultural setting, paradoxically, excluded exchanges between adjacent cultures and, consequently, increased the exceptionalism of Europe, disregarded extra-European factors, and decontextualized early modern science in Europe from its wider global settings. A cross-cultural history of science overcomes such methodological obstacles. It contextualizes the transcendent history of ideas, it crosses the division between print and scribal culture, it avoids teleological notions about the decline and progress of scientific theories, and, finally, it opens up compressed, self-contained cultural histories and allows for a broader and deeper contextual account. Such a prismatic and dialogic approach may reframe central themes in early modern history in general, and with regard to particular questions.

Debates on the historiography of print, for instance, revolve mostly around whether the fixed text represented an agent of transformation or whether social practices determined the use of print technology.[2] Both views imply that

European print culture was incommensurate with the scribal culture of the Near East. Readers of post-Copernican writings, however, inspired a cosmopolitan critique of print culture, which accordingly became a censor, or filter, for ancient sources. Scholars such as Commandino, della Valle, Delmedigo, and Greaves had further argued that printers preferred to exclude many ancient Eastern sources, which therefore remained beyond the print horizon. As a result, practitioners aspired to rescue the, arguably, narrow-minded print culture and traveled eastward in search of seemingly extinct ancient manuscripts.

Actors in the cultural overlay between Europe and the Near East did not see the lack of print culture as an impediment to change but, rather, appreciated the decentralized and less controlled scribal culture in the East. On the one hand, a trip to the East offered the ability to transcend time and to connect, through manuscripts, to the pure wisdom of antiquity and to negotiate current cosmological views with medieval manuscripts in Arabic, Hebrew, Persian, and Latin. On the other hand, local practitioners who encountered printed books (delivered through gift exchanges, the looting of ships, and trade) overturned the order of text production and created handwritten versions of the received printed books. While moving through local scholarly circles, the new cosmology was negotiated and accommodated to existing philosophical and mystical frameworks.

The locality of the Renaissance should also be questioned. Historiography of the Renaissance describes the revival of classical literature as the recovering of lost historical links between European culture and the Greek and Roman civilizations. Renaissance was thus presented as a purely European process, deeply ingrained in Italian society, with no relation to adjacent cultures. Late Renaissance scholars, however, not only thought "that something essential had got lost in the ancient past," to use Frances Yates's words,[3] but developed historical practices of traveling, collecting, and recovering remnants of *prisca theologia* in its homeland—the Near East.

Early modern science, then, did not develop along separate, linear paths, with each culture drawing only on its own "monadic" nature, but rather was cultivated by streams of fresh ideas and objects that came through exchanges at the cultural margins. The principle of the "incommensurability of cultures" would, perhaps, best give way to a more continuative and dialogic approach that sets its eyes not on "cultural centers" but on the hazy yet fertile "cultural margins" that necessarily overlap with other "cultural margins," creating a stimulating, mutually embraced zone where intensive cross-cultural exchanges transpire.

ACKNOWLEDGMENTS

I owe great debts to many individuals. My advisor, Theodore Porter, envisioned the potential of the project from our first meeting. I am especially grateful to him for showing me that historical scholarship requires not only strong evidence, clarity of mind, creativity, and power of imagination, but also a good sense of humor. I was fortunate to have a dissertation committee with each member covering a particular angle of this new niche. Herbert Davidson, Robert Westman, Norton Wise, and Hossein Ziai perfected my training in completely different fields and, thus, set the foundations for my training as a cross-cultural historian of science. Their guidance enriched the scope of my work and rescued it from internal dialogues.

In the course of my work, I exchanged ideas with scholars from adjacent fields. John Christianson, Robert Evans, and Paul Rose gave illuminating comments and suggestions for chapter one and helped in solidifying my argument on the apocalyptic connection in astronomy between the Europeans and the Ottomans. Mario Biagioli and Nick Wilding offered stimulating suggestions on the possible connections to Galileo in the East. David Myers and the late Richard Popkin helped me place chapter three in a larger context of Jewish history and its critical dialogue with Christianity. I had fruitful discussions with Amir Alexander on the connection between voyages, mathematics, and politics in seventeenth-century England. Marry Terrall read most of the chapters and made constructive comments that helped me set the work in the larger context of print culture. George Saliba frequently provided wonderful insights, particularly concerning Arabic astronomy and its circulation in late medieval Europe. Benjamin Elman helped me situate the research in a larger discussion of the world history of science, with special reference to China. Owen Gingerich read the entire manuscript and made many useful comments, and he was particularly helpful in supplying additional documentation for sorting out the intricate publishing strategy of the

French cosmographer Noël Duret. Chapter five started as a seminar paper for Carlo Ginzburg, who first set me on the convoluted path to micro-history.

Zvi Ben-Dor and Eugene Sheppard were roommates, close friends, and colleagues who enriched and stimulated some of the ideas in this book and also supplied strong moral support.

My work was supported by several fellowships, to which I owe a great deal. The Amado Foundation, the International Sephardic Education Foundation, and an International Studies and Overseas Programs fellowship helped me conduct my research overseas. A Clark Library fellowship enabled me to enjoy that library's endlessly fascinating primary sources and the generous help of the librarians. The Milton Fund of Harvard University supported the transformation of my dissertation into a book manuscript. Diana Morse from the Harvard Society of Fellows facilitated the preparation of the manuscript and generously supplied the necessary means to promote its publication.

I am indebted to the museums, archives, and galleries that permitted me to reproduce artwork from their collections. I thank the British Museum, Istanbul University, the Clark Library at the University of California, Los Angeles, the Chester Beatty Library in Dublin, the Houghton Library at Harvard University, the Institute of Hebrew Manuscripts in Jerusalem, and Gallerie dell'Accademia in Venice.

I am lucky to have close friends with whom I could share immature thoughts and first drafts of chapters. Ahmad Alwisha, Courtney Booker, Hillel Eyal, Ming-hui Hu, Minsoo Kang, Margaret Kuo, Kevin Lambert, Nitzan Leibovic, Ofer Nur, and Sandy Sufian were subjected to endless tedious discussions that helped me crystallize my thoughts. During my years in the Harvard Society of Fellows, I was fortunate to have friends and colleagues who sustained a substantial intellectual exchange. The many conversations with Bernard Bailyn, Joshua Blum, Debora Cohen, Peter Galison, Michael Gordin, Anna Henchman, Scott Johnson, Eric Nelson, Vanessa Ryan, Amartya Sen, Andrew Strominger, and Nur Yalman helped me solidify my arguments and expand them to other intellectual fields. Conversations with Ann Blair, Daniela Bleichmar, Jimena Canales, Daniel Margocsy, and Steven Shapin supplied many constructive comments.

The original text went through several cycles of editing. Howard Goodman, friend and colleague, edited the last draft and made many fruitful suggestions. The illuminating responses of the reviewers supplied new research directions and constructive critiques that improved the text significantly. Bob Brugger and his colleagues at the Johns Hopkins University Press helped me escape some convoluted private thinking, making my arguments more straightforward and accessible to readers.

On a more personal level, I am blessed to have friends who gave me moral and emotional support during an extended period of solitude: Mordechai and Tirza Cohen, Charlie Elbaz, Igal Fedida, Paulina Frammer, Allan Gordon, Isaac Halleluiah, Shimon Peretz, Yaacov and Dina Pinto, Gabi Robaz, and Keren Sagiv—I thank them all. While preparing the last draft of the manuscript, I married Atar, and we had our daughters, Michaella and Eleanor. Their endless love and patience enabled me to sort out unseen typos and other embarrassing mistakes in the manuscript. Finally, I want to thank my parents and my extended family, who supported and encouraged me with much love and trust, on good and bad days. Without them my work on this book could not have been completed.

Introduction · *Incommensurable Cultures?*

1. "Tarjama Muqaddimāt Wâhiya fi Ahvāl al-Nujūm zū Zuāba," Hafid Collection, MS 180, Süleymaniye Kütüphanesi, Istanbul.

2. Recent accounts in reaction to the "decline thesis" have intensified the notion of internal dialogue. Such accounts have subverted the attractive notion of a deep rupture, a historical break, and show that after the "Scientific Revolution" in Europe, Islamic science enjoyed a period of reinvigoration. These approaches have created a new genre: the apologetics of Islamic science that stress the lack of contradiction between "true" Islam and "true" science and stress, instead, a chasm between Islam and Western (or non-true) scientific cultures. C. A Qadir, for example, sees a clash of civilizational values between Western science and Islam. He argues that severed from its moral and spiritual foundations, Western scientific culture is lost in a soulless labyrinth of mechanical materialism, where humanism has been lost and life is robot-like. C. A. Qadir, *Philosophy and Science in the Islamic World* (London: Routledge, 1990). In a different vein, Muhammad Mirza and Muhammad Iqbal Siddiqi offer a universalist take: Europe owes a deep debt to Arab scientists. The debt is not for startling discoveries or revolutionary theories but for Muslim scientific methods, some of which traveled to Europe in the late medieval period. Muhammad R. Mirza and Muhammad Iqbal Siddiqi, *Muslim Contribution to Science* (Lahore: Kazi Publications, 1986). However, this approach does not take into consideration non-Muslim practitioners of natural philosophy in the Near East.

3. Historians of Islamic science have argued that Islamic science not only was left behind but lost track of its glory days. Early modern science in the Muslim world seemed to be detached from its own traditions and was more readily identified with Europe's "dark ages." Such scholars give a variety of explanations for the causes of the decline. In the 1960s, Aydin Sayili explained decline as the result of belief in the omnipotence of God. Aydin Mehmed Sayili, "The Causes of the Decline of Scientific Work in Islam," in his *The Observatory in Islam and Its Place in the General History of the Observatory* (Ankara: Türk Tarih Kurumu Basımevi, 1960), 422. Von Grunebaum argued that the problem was to be found in Islamic faith and in the ability of

the orthodox to stifle a once-flowering science. Gustave Edmund von Grunebaum, *Islam: Essays in the Nature and Growth of a Cultural Tradition* (Menasha, WI: American Anthropological Association, 1955), 120. Other historians, such as David King, ascribed the decline to the invasions that plagued the Arab world from the eleventh through the fifteenth centuries. David King, "The Astronomy of the Mamluks," *Isis* 74 (1983): 151. This "declinism" has passed down to contemporary scholarship. In one of the few comparative accounts of the history of science, Toby Huff discusses, in Weberian fashion, the reasons for developments and underdevelopments in natural philosophy in Europe, Islam, and China. Modern European science resulted from the development of a civilizationally based culture that was uniquely humanistic, because it tolerated heretical and innovative ideas. Natural philosophy in the Muslim world, because of the weak role of law and the power of the clergy, was narrowed down to discussions of what is acceptable. Toby Huff, *The Rise of Early Modern Science: Islam, China, and the West* (Cambridge: Cambridge University Press, 1993). See also the critical review by George Saliba, "Seeking the Origins of Modern Science?" *Bulletin of the Royal Institute for Inter-Faith Studies* (*BRIIFS*) 1, no. 2 (1999). To balance these views, I have suggested, elsewhere, that one cannot blame some sort of "internal flaw" in Ottoman intellectual culture, but must look to the real constrictions in daily life—for example, the economic crisis of the religious endowments (*waqf*s) that supported the college system (*madrasah*). Avner Ben-Zaken, "Political Economy and Scientific Activity in the Ottoman Empire," in *The Turks*, ed. H. C. Güzel, C. C. Oğuz, and O. Karatay (Ankara: Yeni Türkiye Publications, 2002), 3:776–94.

4. Since 1957, Otto Neugebauer, Edward Kennedy, Willy Hartner, Noel Swerdlow, and George Saliba, among others, have determined that the mathematical edifice of Copernican astronomy could not have been built, as it was finally built, by using only the mathematical information available in such classical Greek mathematical and astronomical works as Euclid's *Elements* and Ptolemy's *Almagest*. See, for example, N. Swerdlow and O. Neugebauer, *Mathematical Astronomy in Copernicus' De Revolutionibus* (New York: Springer-Verlag, 1985). Recently, this scholarly tradition on the Islamic origins of Copernicus was summarized by Saliba, who followed a well-known assertion that certain theorems of Islamic mathematicians were necessary for Copernicus to establish his heliocentric system; however, none of these theorems were translated into Latin. Saliba presents evidence that there was no need for texts to be fully translated from Arabic into Latin for Copernicus and his contemporaries to make use of them. There were competent astronomers and scientists—contemporaries of Copernicus, or slightly earlier or later than him—who could read the original Arabic sources and make their contents known to their students and colleagues, in the same environment where Copernicus was attempting to reformulate the mathematical foundations of Greek astronomy. George Saliba, *Islamic Science and the Making of the European Renaissance* (Cambridge, MA: MIT Press, 2007), 193–233.

5. The anthropologist James Clifford traces the continuity of processes of cultural and scientific production from the early modern "cabinets of curiosities," which boasted exotic artifacts. James Clifford, *The Predicament of Culture: Twentieth-Century Ethnography, Literature, and Art* (Cambridge, MA: Harvard University Press,

1988), 215–51. For Latour, the crucial activity is science, which created a perception of a "great divide" between other cultures and the modern West, the latter having accumulated traces and specimens and becoming a "center of calculation." Bruno Latour, *We Have Never Been Modern* (Cambridge, MA: Harvard University Press, 1993), 91–129. However, science is not "dark matter" that fills an infinite void. Science is made in a reciprocal process of borrowing and altering.

One • Trading Clocks, Globes, and Captives in the End Time

1. Pierre Gassendi, *Tychonis Brahei, equitis Dani, astronomorum coryphaei, vita* (Paris, 1654).

2. J. L. E. Dreyer, *Tycho Brahe: A Picture of Scientific Life and Work in the Sixteenth Century* (1890) (New York: Dover, 1963).

3. Victor E. Thoren, *The Lord of Uraniborg: A Biography of Tycho Brahe* (Cambridge: Cambridge University Press, 1990).

4. John Christianson, *On Tycho's Island: Tycho Brahe and His Assistants, 1570–1601* (Cambridge: Cambridge University Press, 2000).

5. Sevim Tekeli, "Nasiruddin Takiyüddin ve Tyche Brahenin rasat aletlerinin mukayesesi" [Taqī al-Dīn and Tycho Brahe's Observational Instruments] (Ph.D. diss., Istanbul University, 1958). For a description of the instruments in the Istanbul observatory, see E. Wiedemann, "Definitionen verschiedener Wissenschaften und über diese verfasste Werke," *Physikalisch-Medizinische Sozietät zu Erlangen* 50–51 (1918–19): 26–28.

6. Muammer Dizer, *Takiyüddin* (Ankara: Kültür Bakanlığı, 1990).

7. Aydin Mehmed Sayili, *The Observatory in Islam and Its Place in the General History of the Observatory* (Ankara: Türk Tarih Kurumu Basimevi, 1960). A cross-cultural scholarship has developed that focuses on the substantial intellectual interaction between the Ottoman Eastern Mediterranean and the Latin West, which began in the mid fifteenth century. See, for example, Franz Babinger, *Mehmed the Conqueror and His Time* (Princeton, NJ: Princeton University Press, 1992). On artisans and art, see Nurhan Atasoy and Julian Raby, *The Pottery of Ottoman Turkey* (London: Alexandria Press, 1989); Robert Ousterhout, ed., *Studies on Istanbul and Beyond: The Freely Papers* (Philadelphia: University of Pennsylvania, 2007). On Greek-Arabic and Arabic-Greek translation, see Maria Mavroudi, "Late Byzantium and Exchange with Arabic Writers," in *Byzantium, Faith, and Power (1261–1557): Perspectives on Late Byzantine Art and Culture*, ed. S. T. Brooks (New Haven, CT: Yale University Press, 2007), 62–75. The effects of exchanges on European culture are explored in various works. See, for example, Gerald MacLean, ed., *Re-orienting the Renaissance: Cultural Exchanges with the East* (New York: Palgrave Macmillan, 2005); Jerry Brotton, *Trading Territories: Mapping the Early Modern World* (Ithaca, NY: Cornell University Press, 1998); Lisa Jardine and Jerry Brotton, *Global Interests: Renaissance Art between East and West* (Ithaca, NY: Cornell University Press, 2000).

8. ʿAlāʾ al-Dīn al-Manṣūr, *Shāhinshāhnāma* (ca. 1581), F 1404, Istanbul University Library. Over the past three decades, much of the scholarship on Taqī al-Dīn and his

observatory, especially that by Sayili and his "institutional history" of the Islamic observatories, has relied on *Shāhinshāhnāma*. However, this scholarship has taken the text as self-contained and has produced little on the scientific culture of astronomical observation.

9. See Aydin Sayili, "'Alā' al-Dīn al-Manṣūr's Poems on the Istanbul Observatory," *Türk Tarīh Kurumu Belleten* 79 (1956): 429–84; Sayili, *Observatory in Islam*, 289–305.

10. Sayili, "'Alā' al-Dīn al-Manṣūr's Poems," 445–46.

11. Ibid., emphasis added. For the Persian text and English translation of poems on the observatory, see ibid., 472.

12. Ibid., 473.

13. For instance, the Ottomans did not build their first compass until 1727, based on European models. On examination, it was observed that rather than pointing due north, the needle inclined 11.5 degrees west. See Ekmeleddin İhsanoğlu, "Introduction of Western Science to the Ottoman World: A Case Study of Modern Astronomy (1660–1860)," in *The Transfer of Modern Science and Technology to the Muslim World*, ed. Ekmeleddin İhsanoğlu (Istanbul: Research Centre for Islamic History, Art, and Culture, 1992), 37, 84.

14. *The Murad III Globes: The Property of a Lady to Be Offered as Lot 139 in a Sale of Valuable Travel and Natural History Books, Atlases, Maps, and Important Globes on Wednesday 30 October 1991 . . .* (London: Christie, Manson & Woods, 1991).

15. Schöner did not include a celestial map in his *Opera mathematica*. What he had was a list of stars and their positions. Apparently the map was added in the posthumously published *Opera*.

16. Carlo Cippola, *Clocks and Culture: 1300–1700* (London: Collins, 1967), 87.

17. F. Babinger, "Maometto II conquistatore e l'Italia," *Rivista storica italiana* 63 (1951): 469–505.

18. Otto Kurz, *European Clocks and Watches in the Near East* (London: Warburg Institute, 1975), 30, n. 1.

19. Ibid., 34.

20. Ibid., 47–49.

21. Charles Thornton Forster and F. H. Blackburne Daniell, eds., *The Life and Letters of Ogier Ghiselin De Busbecq, Seigneur of Bousbeque Knight, Imperial Ambassador* (London: C. K. Paul, 1881), 101–2.

22. Taqī al-Dīn Muḥammad Ibn-Ma'ārūf, *Kitāb al-kawākib al-durrīyah fī bānkāmāt al-dawrīyah* [The Revolving Planets and the Revolving Clocks], Mīqāt Collection, MS 557/1, Dār al-Kuttub, Cairo.

23. Taqī al-Dīn Muḥammad Ibn-Ma'ārūf, "Baqiyyat al-ṭullāb ilā 'ilm al-ḥisāb" (1578), Carullah Collection, MS 1454, Süleymaniye Manuscript Library, Istanbul. I thank Ihsan Fazlioglu for informing me of the existence of this manuscript in the archive.

24. Taqī al-Dīn Muḥammad Ibn-Ma'ārūf, *Kitāb al-ṭuruq al-samiyyah fī al-ālāt al-rūḥānīyyah*, facsimile in *Taqī al-Dīn wa al-handasah al-mīkānīkiyyah*, ed. Aḥmad Yūsuf al-Ḥasan (Aleppo: University of Aleppo, 1976).

25. The Syrian scholar Aḥmad Yūsuf al-Ḥasan has surveyed different accounts of Taqī al-Dīn's identity and places his birth and education in Syria and the sources of his knowledge in mechanics, alchemy, and talismans as coming from an early education at schools in Nablus and Damascus. Ibid., 18–19.

26. Ibid., 19.

27. Ibid., 65–68.

28. Ibid., 3. See also Taqī al-Dīn, *Kitāb al-kawākib al-durrīyah fī bānkāmāt al-dawrīyah*, v.4b, v.5a, as cited in Sevim Tekeli,*16'ıncı asırda Osmanlılarda saat ve Takiyüddin'in Mekanik saat konstrüksüyonuna dair en parlak yıldızlar adlı eseri: The Clocks in Ottoman Empire in 16th Century and Taqī al-Dīn's "The Brightest Stars for the Construction of the Mechanical Clock"* (Ankara: Ankara Universitesi Basimevi, 1966), 215–323.

29. For a contemporary biography of ʿAlī Pasha, see Ahmad Shalbī Ibn ʿAbd al-Ghanī, *Awḍaḥ al-ishārāt fiman wallī Miṣr al-Qāhira min al-wuzārāʾ wa-al-bāshāt*, ed. Fu'ād Muḥammad al-Māwī (Cairo: Kuliyyat al-Lughah al-ʾArabiyyah, Jāma'at al-Azhar, 1977), 148.

30. Taqī al-Dīn, *Kitāb al-ṭuruq al-samiyyah fī al-ālāt al-rūḥānīyyah*, 3–5.

31. Taqī al-Dīn, *Kitāb al-kawākib al-durrīyah fī bānkāmāt al-dawrīyah*, v.4b.

32. Taqī al-Dīn Muḥammad Ibn-Ma'ārūf, *Sidrat al-muntah al-afkār fi malkūt al-falak al-dawār al-zīj al-Shāhinshāhī*, MS 2930, Nuruosmaniye Library, Süleymaniye Kütüphanesi, Istanbul, pp. 85b, 86b–87.

33. Taqī al-Dīn, *Kitāb al-kawākib al-durrīyah fī bānkāmāt al-dawrīyah*, v.4b.

34. Ibid., L.3a.

35. Ibid., v.4a.

36. Donald Hill, *Arabic Water-Clocks* (Aleppo: University of Aleppo, 1981), 1.

37. See The Venerable Bede, *De temporum ratione*, in *Bedae opera de temporibus*, ed. C. W. Jones (Cambridge, MA: Mediaeval Academy of America, 1943), 195; for further reading see Gerhard Dohrn-van Rossum, *History of the Hour: Clocks and Modern Temporal Orders* (Chicago: Chicago University Press, 1996), 20.

38. Ismā'īl ibn ʿUmar Ibn Kathīr, *Kitāb al-nihāyah, aw al-fitan wa-al-malāḥim* [The Book of the End: Great Trials and Tribulations], ed. Ṭāhā Muḥammad al-Zaynī (Cairo: Dār al-Kutub al-Ḥadīthah, 1969), 1:55.

39. On this tradition of the Last Hour, see Khālid ibn Nāṣir ibn Sa'īd Ghāmidī, *Ashrāṭ al-sā'ah fī mustanad al-imām Aḥmad wa-zawā'id al-ṣaḥīḥayn: jama'an wa-takhrījan wa-sharḥan wa-dirasah* (Jiddah: Dār al-Andalus al-Khaḍrā', 1999), 2 vols. For secondary literature, see ʿIṣām Sayyid, *al-Fitan wa-ʾalāmāt ākhir al-zamān lil-Imāmayn al-Bukhārī wa-Muslim* (al-Jīzah: Maktabat al-Nāfidhah, 2003); Sandra Campbell, "It Must Be the End of Time: Apocalyptic *Aḥādīth* as a Record of the Islamic Community's Reactions to the Turbulent First Centuries," *Medieval Jewish, Christian, and Muslim Culture Encounters in Confluence and Dialogue* 4, no. 3 (1998): 178–87; Suliman Bashear, "Muslim Apocalypses and the Hour: A Case Study in Traditional Representation," in *Israel Oriental Studies XII*, ed. Joel Kramer (Leiden: Brill, 1993); Norman Brown, *Apocalypse and/or Metamorphosis* (Berkeley and Los Angeles: University of California Press, 1991), 69–95.

40. Ibn Kathīr, *al-nihāyah, aw al-Fitan wa-al-Malāḥim*, 1:6.

41. Taqī al-Dīn, *Kitāb al-kawākib al-durrīyah fī bānkāmāt al-dawrīyah*, v.1b.

42. Ibid., v.4b.

43. 'Alā' al-Dīn al-Manṣūr, *Shāhinshāhnāma*, in Sayili, "'Alā' al-Dīn al-Manṣūr's Poems," 472–73.

44. See, for example, Sayili, *Observatory in Islam*, 221, 240.

45. Taqī al-Dīn, *Sidrat al-muntah al-afkār fi malkūt al-falak al-dawār al-zīj al-Shāhinshāhī*, 85b, 86b–87a.

46. On the close cultural and economic connections of the Salonika community with Italy, see Meir Benayahu, *HaYahasim sheBen yehude Yavan ve yehude Italia* [The Relations between Greek and Italian Jewry] (Tel-Aviv: Diaspora Research Institute, 1980).

47. See J. O. Leibowitz, *Amatus Lusitanus (1511–1568) è Salonique* (Rome: Arti grafiche e cossidente, 1970).

48. See Ekmeleddin İhsanoğlu, *Büyük Cihad'dan Frenk fodulluğuna* (Istanbul: İletişim, 1996), 106–9; Ekmeleddin İhsanoğlu, *Osmanlı astronomi literatürü tarihi* [History of Astronomy Literature during the Ottoman Period] (Istanbul: İslâm Tarih, Sanat ve Kültür Araştırma Merkezi, 1997), 1:328–29; J. H. Mordtmann, "Das Observatorium des Taqī en-din zu Pera," *Der Islam* 13 (1923): 92.

49. David A. Rekanati, *Zikhron Saloniki: gedulatah ve-hurbanah shel Yerushalayim de-Balkan, ha-'orekh* (Tel Aviv: ha-Va'ad le-hotsa'at sefer Kehilat Saloniki, 1971); Michael Molco, *Beit ha'Almin shel Saloniki* [The Cemetery of Saloniki] (Tel Aviv: Hotsaat Makhon le-heker Yahadut Saloniki, 1974).

50. HaLevi's comment is in the introduction to Joseph Solomon Delmedigo, *Sefer Elim* (Amsterdam: Menasheh ben Yisrael, 1628), 4b.

51. David Conforte, *Kore haDorot* [The Reader for the Generations] (Berlin: Abraham ben Asher, 1846), 39

52. Thomas Bricot, *Toldot HaAdam*, trans. David Ben-Shushan, MS 5475, Jewish Theological Seminary, New York.

53. Ibid., 59b.

54. Ibid., 76b. We have some difficulty in finding indications of why Ben-Shushan chose to translate Bricot's commentary on Aristotle, because the manuscript does not contain the first or last pages, where we would expect some sort of introduction or comments by the translator. We have few other writings of Ben-Shushan. There is an indication that he wrote a work titled *Biet habhira* [House of Will], but it did not survive. Remnants of the introduction to this work, however, are attached elsewhere (untitled MSS, B 267, 4a–9b, St. Petersburg–Institute of Oriental Studies of the Russian Academy, St. Petersburg). These remnants indicate that the work dealt with the question of determinism and free will, especially in the light of Ben-Shushan's more famous work, *Peirush Kohelet* [Commentary on Ecclesiastes] (MS Heb. 4°619, Jewish National and University Library, Jerusalem), in which he addresses the question of the status of the human spirit (*rūah, rūh, pneuma*). Langermann argues that Ben-Shushan was contemplating questions of free will and determinism. Spirit is the substrate of the soul, but, as such, is it divine like the soul, so that it too ascends after

death? Or is it purely material and hence perishable? After reviewing a number of medieval Islamic sources, Ben-Shushan decides in favor of the view of Ibn Tufayl, who, in his "philosophical romance" *Hayy Ibn Yaqzān*, declares the spirit to be divine. Among the interpretations that Ben-Shushan rejects is that of "the authors of the Zohar." However, his critique is conducted entirely in a scientific idiom, without any polemical overtones. This is instructive insofar as it illustrates that cabalists and natural philosophers of the period engaged, on the whole, in a constructive discourse based on shared concepts. The texts that I study in this book testify to the endurance of Andalusian Jewish learning even after the expulsion of 1492. Indeed, perhaps Giordano Bruno mined some of the same sources used by Ben-Shushan. See Tzvi Langermann, "David Ibn Shoshan on Spirit and Soul," *European Journal of Jewish Studies* 1, no. 1 (2007): 63–86.

55. Taqī al-Dīn, *Sidrat al-muntah*, 85b, 86b–87a.

56. Abraham Ben-Eliezer HaLevi, *Ma'amar Meshare Qitrin* (Constantinople, 1510), ed. Gershom G. Scholem (Jerusalem: Jewish National Library Press, 1977). In this book, HaLevi—who was expelled from Spain and moved to Italy and then to Ottoman Salonika—predicted that the apocalypse would occur between 1530 and 1540, the final event being the destruction of Rome by the Ottomans. He based his predictions on knowledge about great planetary conjunctions, which he learned from his brother-in-law, Avraham Zacut. According to Zacut and HaLevi, the last conjunction of this kind had occurred when Joshua conquered the land of Canaan, and therefore it signified the last redemption of the Jews (ibid., 26). Later, the same argument was used in reverse by two famous figures. In the 1530s, David Reuveni declared that Shlomo Molcho (d. 1532) was the messiah. Based on Molcho's apocalyptic visions, Reuveni and Molcho suggested to the pope that a Jewish-Christian alliance should be formed against the "Turks." They argued that only the doom of the "Turks" would lead to the liberation of the Holy Land and the construction of the Temple. Julius Voos, *David Reubeni und Salomo Molcho; ein Beitrag zur Geschichte der messianischen* (Berlin: Buchdruckerei Michel, 1933). In the eighteenth century, the same argument was used again by John Collet. According to Collet's treatise on the redemption of the Jews, the coming of the messiah would occur only with the fall of the "Turks." John Collet, *A Treatise of the Future Restoration of the Jews and Israelites to Their Land: With Some Account of the Goodness of the Country, and Their Happy Condition There, Till They Shall Be Invaded by the Turks: With Their Deliverance from All Their Enemies, When the Messiah Will Establish His Kingdom at Jerusalem, and Bring in the Last Glorious Ages* (London: Printed for J. Highmore, M. Cooper, and G. Freer, 1774).

57. An apocalyptic work from the first century, Zerubavel Ben-Shealtiel's *Hazon Zerubavel* [Vision of Zeruvavel], was printed for the first time in 1503 in the Hebrew press of Constantinople.

58. Benayahu, *HaYahasim sheBen yehude Yavan ve yehude Italia*, 230–36.

59. Almosnino translated two astronomical pieces from Latin to Hebrew. The first (*Beit Elohim*) was a thirteenth-century treatise, *Sphere mundi*, by Johannes Sacrobosco, John of Holywood, who mostly relied on Greek and Muslim astronomers; the second (*Shaar HaShammyyim*) was a fifteenth-century work, *Theoricae novae*

planetarum, by George Peurbach. Moshe Almosnino, "*Beit Elohim: Perush Kadur Ha'Olam Le Yohanes Scarobosco* [The House of God: Commentary on the Globe of the Universe of Johannes Sacrobosco], Schoenberg Collection, Ljs 42, University of Pennsylvania, Philadelphia; Moshe Almosnino, *Shaar HaShammyyim* [The Gate to Heaven; a translation of Peurbach's new theory of the planets], Evr. II A 15, Russian National Library, St. Petersburg.

60. Nasi had banking connections with the Spanish and French courts; he fled to Istanbul when his businesses could not remain in Christian Europe. See Norman Rosenblatt, "Joseph Nasi: Court Favorite of Selim II" (Ph.D. diss., University of Pennsylvania, 1957), 99–105.

61. Salomon Schweigger, *Ein newe Reyssbeschreibung auss Teutschland nach Constantinopel und Jerusalem*, ed. Rudolf Neck Wein (Graz: Akademische Druck- u. Verlagsanstalt, 1964), 90–91. I thank Christina Kurtz for help in translating this text.

62. Taqī al-Dīn's having Christian captive-artisans was not unusual. Braudel asserts that sixteenth-century Mediterranean artisans consisted of many races, rarely native to the area, and that Istanbul was "where manufacture was often in the hands of immigrants, Christian prisoners who at Constantinople and elsewhere frequently became master of craftsmen." Most artisans in small industries were members of the Jewish community of Salonika. Fernand Braudel, *The Mediterranean and the Mediterranean World in the Age of Philip II* (New York: Harper & Row, 1966), 1:436.

63. Schweigger, *Ein newe Reyssbeschreibung*, 90–91, emphasis added. This account by Schweigger was mentioned at the beginning of the twentieth century in Mordtmann, "Das Observatorium," 89–92. It was ignored by Middle Eastern scholars interested in preserving an Islamic authenticity for Taqī al-Dīn's work. See Dizer, *Takiyüddin*, 38–40, 297; Tekeli, *16'ıncı asırda Osmanlılarda saat*, 124–35.

64. Taqī al-Dīn noted, by hand, on his copy of the Arabic translation of the *Almagest* that he read about Ptolemy in the dictionary of the multilingual Ambrogio Calipino (d. 1511). *Almagest* (in Arabic), Arabic no. 7116, Bibliothèque nationale de Tunisie, Tunisia. I thank George Saliba, who found this manuscript, for sharing the information. See also George Saliba, "The World of Islam and Renaissance Science and Technology," in *The Arts of Fire: Islamic Influences on Glass and Ceramics of the Italian Renaissance*, ed. Catherine Hess (Los Angeles: J. Paul Getty Museum, 2004), 69–71.

65. Salvatore Bono, *Corsari nel Mediterraneo: cristiani e musulmani fra guerra, schiavitù e commercio* (Milan: A. Mondadori, 1993). See also Gustavo Valente, *Calabria Calabresi e Turcheschi nei secoli della pirateria (1400–1800)* (Frama: Chiaravella Centrale, 1973).

66. For the memoirs of Muṣṭafā Efendi, see F. Schmucker, "Die maltesischen Gefangenschaftserinnerungen eines türkischen Kadi von 1599," *Archivum Ottomanicum* 2 (1970): 191–251.

67. Eliezer Bashan, *Shviyya ve pedut* [Captivity and Redemption] (Tel Aviv: Bar-Ilan University, 1980), 134.

68. Orhan Pamuk, in his novel *The White Castle*, transforms an actual autobiographical manuscript of a European captive. The overall sense of the novel, which

is quite accurate, is that the Ottomans were aware of European technological and scientific advantages and used captives as sources of knowledge. Orhan Pamuk, *The White Castle* (New York: Braziller, 1991).

69. Nabil Matar gives us examples of journals from the seventeenth century that were written by Muslims who went to Europe either to ransom captives or to learn about the new maritime explorations and the discoveries coming from the New World. Nabil Matar, ed. and trans., *In the Lands of the Christians: Arabic Travel Writing in the Seventeenth Century* (New York: Routledge, 2003).

70. See Karl Dannenfeld, "The Humanists' Knowledge of Arabic," *Studies in the Renaissance* 2 (1955): 96–117. For a bibliography of all pre-1919 works printed in Arabic, see Yūsuf Ilyān Sarkīs, *Mu'jam al-maṭbū'āt al-'Arabīyah wa-al-mu'arrabah, wa-huwa shāmil li-asmā' al-kutub al-maṭbū'ah fī al-aqṭar al-sharqiyyah wa-al-gharbiyyah, ma'a dhikr asmā' mu'allifīhā wa-lum'ah min tarjamatihim. wa-Dhalik min yawm ẓuhūr al-ṭabā'ah ilā nihāyat al-sanah al-hijriyyah 1339 al-muwāfiqāh li-sanat 1919 mīlādiyyah* (Cairo: Sarkīs, 1928). See also a series of articles by L. Cheikho, "Tārīkh fann al-ṭibā'a fi al-mashriq" [History of the Art of Book-Printing in the East], *al-Mashriq*, 1900–1902, pp. 3–5. My discussion here is based on Johannes Pedersed, *The Arabic Book*, trans. Geoffrey French (Princeton, NJ: Princeton University Press, 1984), 131–41; and Samīr 'Aṭā Allāh, *Tārīkh wa-fann ṣinā'at al-kitāb* (Beirut: Dār 'Aṭā Allāh, 1993), 124–26.

71. Leo Africanus, *A History and Description of Africa*, trans. John Pory (1600) (London: Hakluyt Society, 1896). Leo also wrote an Arabic-Spanish vocabulary for the instruction of his pupil Jacob Mantino, the celebrated Jewish physician. Harwig Derenbourg, "Leon Africain et Jacob Mantino," *Revue des Études Juives* 7 (1883): 283–85. Leo also taught Arabic in Rome, and the humanist Cardinal Gilles of Viterbo (Aegidius) was among his first pupils. Johann Albrecht Widmanstetter, *Liber sacrosancti evangelii de Iesv Christo* . . . (Vienna: M. Zymmerman, 1562), fol. 12b. See also Natalie Zemon Davis, *Trickster Travels: A Sixteenth-Century Muslim between Worlds* (New York: Hill and Wang, 2006).

72. Jerry Brotton, "Printing the World," in *Books and the Sciences in History*, ed. Nicholas Jardine and Marina Frasca-Spada (Cambridge: Cambridge University Press, 2000), 35–48.

73. For a full account of Muslim presence in early modern Italy, see Salvatore Bono, *Schiavi Muslamani nell'Italia moderna Galeotti, vu' cumpra, domestici* (Naples: Edizioni Scientifiche Italiane, 1999).

74. See Paul Lawrence Rose, "Humanist Libraries and Renaissance Mathematics: The Italian Libraries of the Quattrocento," *Studies in the Renaissance* 20 (1973): 46–105.

75. In his translation of Ptolemy, Commandino dedicated the commentary to Ranuccio Farnese and took that occasion to explain how contemporary mathematicians complained of the extreme difficulty in reading Ptolemy's *Planisphaerium*. Paul Lawrence Rose, *The Italian Renaissance of Mathematics: Studies on Humanists and Mathematicians from Petrarch to Galileo* (Geneva: Librairie Droz, 1975), 197.

76. The extant treatise was merely a Latin translation of the ninth-century Arabic

version by Masalma of Córdoba—or Messala, as Commandino referred to him. A certain Balthasar Turrius Metinensis had persuaded Commandino to read through the book and try to render it intelligible. Rose, *Italian Renaissance of Mathematics*, 197. Commandino also recovered Greek mathematics through Arabic texts, including Euclid. In 1563, John Dee visited Commandino, bringing with him a Latin manuscript titled *De superficierum divisionibus*, essentially a version of an Arabic treatise by Muhammad al-Baghdādī (d. 1037) on the division of rectilinear plane figures. Commandino held on to the text for a long time and published it only in 1570, as Muhammad al-Baghdadi, *De superficierum divisionibus liber; Federici Commandini de eadem re libellus* (Pesaro: Concordia, 1570). Dee's letter (probably early 1560s) to Commandino that prefaced the volume conjectured that Muhammad al-Baghdādī's treatise is actually the lost *Liber divisionum* of Euclid. However, Commandino ignored this in his dedicatory letter to Prince Francesco Maria II of Urbino. See Rose, *Italian Renaissance of Mathematics*, 200.

77. Rose, *Italian Renaissance of Mathematics*, 213–14.

78. Bernardino Baldi, *Le vite de' matematici: edizione annotata e commentata della parte medievale e rinascimentale* (Milan: Franco Angeli, 1998).

79. Tycho Brahe, *Tychonis Brahe Dani: opera omnia*, ed. J. L. E. Dreyer (Copenhagen: Gyldendal, 1913–29), 1:135 and 10:13. Dreyer mentions the existence in the Hofbibliothek in Vienna of a Tycho pamphlet, "De Eclipsi Lunari, 1573, Mense Decembri," apparently intended to be printed, which discusses the eclipse of 1566 and the prediction of the sultan's death. Dreyer, *Tycho Brahe*, 26.

80. Tycho Brahe, *Astronomiae instauratae progymnasmatum pars tertia*, in Brahe, *Tychonis Brahe Dani: opera omnia*, 3:315–19.

81. See John Christianson, "Tycho Brahe's German Treatise on the Comet of 1577: A Study in Science and Politics," *Isis* 70 (1979): 138.

82. For Tycho's references to the chaotic war between good and evil, or Gog and Magog, see Brahe, *Tychonis Brahe Dani: opera omnia*, 3:314 and 8:417.

83. *Predictions of the Sudden and Total Destruction of the Turkish Empire and Religion of Mahomet: According to the Opinions of the Lord Tycho Brahe of Denmark, and Many Others of the Best Astronomers of This Later Age, Collected and Humbly Dedicated to All Christendom by a Lover of Christianity* (London: Walter Davis, 1684).

84. All quotations from ibid., 2–4.

85. Christianson, *On Tycho's Island*, 170.

86. Dreyer, *Tycho Brahe*, 262–63.

87. Christianson, *On Tycho's Island*, 69, n. 21.

88. R. J .W. Evans, *Rudolf II and His World: A Study in Intellectual History, 1576–1612* (New York: Oxford University Press, 1973), 152. Habsburg courtiers such as Hagecius had an active part in the cosmological polemic in the 1570s. In his treatises, Hagecius was ultimately concerned with the role of celestial events as portents and as manifestations of an organic universe.

89. Ibid., 54.

90. For instance, in 1596, Nicolas Reymers (Raimarus Ursus), court astrologer to Rudolf II, published an apocalyptic account entitled *Chronological, Certain, and*

Irrefutable Proof, from the Holy Scripture and Fathers; That the World Will Perish and the Last Days Will Come within 77 Years (Nuremberg, 1606).

91. See Karl Vocelka, *Die politische Propaganda Kaiser Rudolfs II: 1576–1612* (Vienna: Verlag der Österreichischen Akademie der Wissenschaften, 1981), 219–301. Rudolf II's reign was preoccupied with the popularization of anti-Ottoman sentiments in order to emphasize Rudolf's apocalyptic role as the savior of Europe. Propaganda was spread through pamphlets, coins, medallions, and sermons (with the Turks presented as the punishment of God). In these sources we find references to emblems, astrology, Jupiter, and Saturn. (The emblems represented Rudolf as defeating the Ottomans.) Rudolf's chamber was filled with iconography of Europe, Mars, and Jupiter (ibid., 214). One painting shows Mercury and Jupiter in the sky, as well as Fama and a figure (Rudolf) with a crown (292). In another painting we find the symbols of the emperor, a lion and an eagle, fighting with a dragon. The planet, Saturn, at the top of the painting symbolizes Rudolf's victory as corresponding to God's will (295). I thank Simone Kussatz for her help in translating Vocelka's text.

92. See Salomon Schweigger, *Der Turken Alcoran, Religion und Aberglauben* (Nuremberg, 1616, 1623).

93. Václav Budovec, *Anti-Alkoran, to jest mocní a nepřemožení důvodové toho, ze Al-Koran Turecký z d'ábla pošel* (Prague, 1614). Although not published until 1614, the *Anti-Alkoran* was written in the 1590s, together with some less comprehensive tracts, and these were already circulating in manuscript form before that time—a copy was made of them in 1608 for Peter Vok Rožmberk, leader of the Czech contingent sent against the Ottomans in 1594. See R. J .W. Evans, "Bohemia: The Emperor, and the Porte, 1550–1600," *Oxford Slavonic Papers* 3 (1970): 101.

94. See Evans, *Rudolf II and His World*, 109. Palaeologus, at the time an ex-Dominican theologian, created an influential circle in Prague, including Hagecius (Tycho's close friend). The circle aimed to undermine Ottoman Istanbul by tying the cultural underpinning of Prague with that of Greek Constantinople.

95. Abū-Ma'shar, *Kitāb al-milal wa al-duwal*, ed. K. Yamamoto and C. Burnett (Leiden: Brill, 2000), 1:63.

96. This image of Selim as drunkard was stressed to the extent that a contemporary poet, Bākī, wrote that "the late Sultan Selim of angelic morals had said [of himself] 'I am that one sound of nature who inclines to ruddy wine,'" and was accused of using derogatory images of Sultan Selim II. See Walter Andrews, Najaat Black, and Mehmet Kalpakli, eds., *Ottoman Lyric and Poetry: An Anthology* (Austin: University of Texas Press, 1997), 240–41. Selim II's drinking was a sign of the apocalypse, according to the Signs of the Last Hour ("Drinking and fornication will increase heavily"). The contemporary historian 'Ali Muṣṭafā wrote that decline became even more apparent under Selim II. See Cornell Fleischer, *Bureaucrat and Intellectual in the Ottoman Empire: The Historian Mustafa 'Ali* (Princeton, NJ: Princeton University Press, 1986), 259.

97. See Muṣṭafā Soykut, *Image of the "Turk" in Italy: A History of the "Other" in Early Modern Europe: 1453–1683* (Berlin: Klaus Schwarz Verlag, 2001), 62–65. In Jewish sources from the Balkans, the battle of Lepanto was presented as shaking the foun-

dations of communal life. Eliezer Bashan, *Shviyya ve Pedut* [Captivity and Ransom] (Ph.D. diss., Bar-Ilan University, 1980), 138, n. 9.

98. For a summary of Mevlānā ʿIsā, see *Cami-ül Meknunat* [*The Compendium of Hidden Things*], in Fleischer, *Bureaucrat and Intellectual*, 165.

99. Ibid., 295.

100. Ibid., 268.

101. Lazaro Soranzo, *L'Othomanno* (Ferrara: Vittorio Baldini, Stamatore Camerale, 1598), 45.

102. Fleischer, *Bureaucrat and Intellectual*, 205. For the works of Muṣṭafā ʿAlī, see Andreas Tietze, *Muṣṭafā ʿĀlī's Counsel for the Sultans of 1581* (Vienna: Verlag der Österreichischen Akademie der Weissenschaften, 1979), 2 vols.; Andreas Tietze, *Muṣṭafā ʿĀlī's Description of Cairo of 1599* (Vienna: Verlag der Österreichischen Akademie der Weissenschaften, 1975); Jan Schmidt, *Muṣṭafā ʿĀlī Künhü'l-Aḥbār and Its Preface According to the Leiden Manuscript* (Istanbul: Nederlands Historisch-Archaeologisch Instituut te Istanbul, 1987).

103. Fleischer, *Bureaucrat and Intellectual*, 301.

104. See Kenneth M. Setton, *Western Hostility to Islam and Prophecies of Turkish Doom* (Philadelphia: American Philosophical Society, 1992), 7–8.

105. See Abu-Ma'shar, *Kitāb al-milal*, 1:115.

106. al-Manṣūr, *Shāhinshāhnāma*, in Sayili, "ʿAlā' al-Dīn al-Manṣūr's Poems," 472.

107. Ibid., 472–73.

108. See Avner Ben-Zaken, "Recent Currents in the Study of Ottoman-Egypt Historiography, with Remarks about the Role of the History of Natural Philosophy and Science," *Journal of Semitic Studies* 49, no. 2 (2004): 303–28.

109. Aḥmad Shalabī Ibn ʿAbd al-Ghanī, *Awḍaḥ al-ishārāt fīman tawallá Miṣr al-Qāhirah min al-wuzarā' wa-al-bāshāt: al-mulaqqab bi-al-tārīkh al-'aynī* (Cairo: Tawzī' Maktabat al-Khānjī, 1978), 140.

110. Muḥammad ibn Muḥammad Murtaḍá al-Zabīdī, *Tāj al-'arūs min jawāhir al-Qāmūs*, V:206f, cited in Stefan Reichmuth, *The World of Murtada al-Zabidi (1732–91): Life, Networks, and Writings* (Oxford: Oxbow Books, 2008), chap. 4.

111. Yosef ben Yitzhak Sambari, *Sefer Divrei Yosef*, ed. Shimon Shtober (Jerusalem: Ben-Zvi Institute, 1994), 78a.

112. Ibid., 475.

113. Ibid., 479. Pleiades is an open star cluster located within Taurus; myths related to Pleiades often represented lost people. Temples were erected and aligned with these stars, and some cultures considered them the center of the universe and the destination of the soul.

114. Sayili, "ʿAlā' al-Dīn al-Manṣūr's Poems," 479–480.

115. Ibid., 480.

116. Abū-Ma'shar, *Kitāb al-milal wa al-duwal*, 1:307.

117. Sayili, "ʿAlā' al-Dīn al-Manṣūr's Poems," 480, 481.

118. Ibid, 481.

119. Ibid., 472.

120. See Robert E. Stout, "The Sûr-I-Hümâyun: A Study of Ottoman Pageantry and Entertainment" (Ph.D. diss., Ohio State University, 1966), 207–9.

121. Ibid., 211.

122. Sayili, "'Alā' al-Dīn al-Manṣūr's Poems," 483.

123. Ibid., 483.

Two • Exchanging Heliocentrism for Ur-Text

1. Pietro della Valle, "Letter to Zayyn al-Dīn al-Lārī," Persian Collection, MS 9, Vatican Library. The letter is briefly mentioned in some secondary sources as evidence for cross-cultural exchange, yet there has been no attempt to contextualize it. For instance, Sayili Aydin briefly describes the astronomical content of the letter without paying any attention to the significance of, for example, the discussions on biblical passages. Sayili Aydin, *Tycho Brahe Sistemi Hakkinda xvii. Asir Baslarina Ait Farça Bir Yazma: An Early Seventeenth Century Persian Manuscript on the Tychonic System* (Ankara: Dil ve Tarih-Cografya Fakültesi, 1958), 79–87.

2. Pietro della Valle, *De viaggi di Pietro della Valle il pellegrino: descritti da lui medesimo in lettere familiari all'erudito suo amico Mario Schipano* (Rome, 1650). I occasionally consulted extracts of della Valle's journal in English in Wilfrid Blunt, *Pietro's Pilgrimage: A Journey to India and Back at the Beginning of the Seventeenth Century* (London: James Barrie, 1953).

3. Giovanni Pietro Bellori, *Vita di Pietro della Valle il pellegrino* (Rome, 1662). His collection of artists' biographies was published later. Bellori, *Le vite de' pittori, scvltori et architetti moderni* (Rome, 1672).

4. For a recent survey of the historiography of the Galilean affair, see Maurice Finocchiaro, "Science, Religion, and the Historiography of the Galileo Affair: On the Undesirability of Oversimplication," *Osiris*, 2nd ser., 16 (2001): 114–32.

5. Mario Biagioli, *Galileo, Courtier: The Practice of Science in the Culture of Absolutism* (Chicago: University of Chicago Press, 1993).

6. See Sonja Brentjes, "Western European Travelers in the Ottoman Empire and Their Scholarly Endeavors (Sixteenth–Eighteenth Century)," in *The Turks* (Ankara: Yeni Turkiye Publication, 2002), 3:795–803. For literature on travelers, see Boies Penrose, *Travel and Discovery in the Renaissance, 1420–1620* (Cambridge, MA: Harvard University Press, 1953); Jonathan Haynes, *The Humanist as a Traveler* (London: Associated University Press, 1986). On specific literature that treated della Valle as a humanist traveler, see Anthony George Bull, *The Pilgrim: The Travels of Pietro della Valle* (London: Hutchinson, 1990); and Blunt, *Pietro's Pilgrimage.*

7. Della Valle, "Letter to al-Lārī," 1. I thank Isaac Haleluya for his help in managing the problems with Persian linguistic corruptions in this letter.

8. Ibid., 4–5.

9. For an extensive study on Borrus, see Domingues Maurício Gomes dos Santos, "Vicissitudes da obra do Pe. Cristóvã Borri," *Anais da Academia Portuguesa de Historia*, 2nd ser., 3 (1951): 117–50.

10. Borrus's work was reprinted in 1940 by the Portuguese government as part of an effort to revive the imperial heritage; Borrus's stay in Goa would have made him part of that heritage. Cristóvão Bruno, *Arte de navegar* (1628) (Lisbon: República Portuguesa, Ministério Das Colónias, 1940). Borrus was also known to be an expert on the history of Chinese and Vietnamese uses and adaptations of cannon technology. See Li Tana, *Nguyen Cochinchina: Southern Vietnam in the Seventeenth and Eighteenth Centuries*, Cornell University, Southeast Asia Program Publications, Studies on Southeast Asia, no. 23 (Ithaca, NY: Cornell University 1998), 44.

11. See Angelo Mercati, "Notizie sul gesuita Cristoforo Borri e su sue 'Inventioni' da carte Finora sconosciute di Pietro della Valle il pellegrino," *Pontificia academia scientiarvm acta* 15, no. 3 (1951–53): 25–46.

12. Norman J. W. Thrower, "Edmond Halley as a Thematic Geo-Cartographer," *Annals of the Association of American Geographers* 59, no. 4 (1969): 664.

13. Cristoforo Borri [Christopher Borrus], *Collecta astronomica: ex doctrina P. Christophori Borri, mediolanensis, ex Societate Iesu; Detribuscaelis. Aereo, Sydereo, Empyreo; Issu, et studio Domini D. Gregorii de Castelbranco Comitis Villae Nouae, Sorteliae, & Goesiae domus dynastae, Regij corporis Cnstodi maximo, &c. Opus sane mathematicum, philosophicum, & theologicum, sive scrpturarium. Superiorum permissu* (Lisbon, 1631).

14. Borrus takes the time to describe the telescope in this way: "[It] finally was perfected in all its measurements for common [widespread] use by Galileo Galilei of Florence." Borri, *Collecta astronomica*, 135–36.

15. Ibid., 141.

16. See Cristoforo Borri, *Relatione della nuova missione delli PP. della Compagnia di Giesu, al regno della Cocincina; An Account of Cochin-China: The First Treatise of the Temporal State of That Kingdom and Second of What Concerns the Spiritual*, trans. Robert Ashley (1633) (London, 1704), 803–4.

17. Ibid., 828–29.

18. Della Valle, *De viaggi*, 2:326–27.

19. Ibid., 2:326.

20. Ibid.

21. Modern studies mention al-Lārī only through the della Valle connection. See Laurence Lockhart, "European Contacts with Persia, 1350–1736," and E. S. Kennedy, "The Exact Sciences in Timurid Iran," both in *The Cambridge History of Iran: Volume Six, The Timurid and Safavid Periods*, ed. Peter Jackson and Laurence Lockhart (Cambridge: Cambridge University Press, 1986), 394–95, 568–80. Kennedy's description mentions al-Lārī based on information from della Valle's *De viaggi*.

22. See Agha Buzurg, *al-Dharī'ah ilá taṣānīf al-Shī'ah, ta'līf Muḥammad al-shahīr bi-al-Shaykh Āghā Buzurg al-Ṭihrānī* (al-Najaf: Matba'at al-Gharri, 1936); Agha Buzurg, *Tabaqat a'lam al-Shi'ah* (Beirut: Dar al-Kitab al-'Arabi, 1971); Ahmad Munzavi, *Fihristvarah-i kitabha-yi Farsi* (Tehran: Anjuman-i Asar va Mafakhir-i Farhangi, 1995).

23. See, for example, two catalogues of Islamic manuscripts on natural philosophy: *Osmanli Astronomi Literaturu Tarihi* [History of Astronomic Literature during the Ottoman Period] (Istanbul: IRICICA, 1997); and David King, ed., *Fihris al-makh*

Ṭū Ṭāt al-ʿilmīyah al-maḥfūẓah bi-Dār al-Kutub al-Miṣrīyah [Catalogue of the Scientific Manuscripts in the Egyptian National Library] (Cairo: al-Hay'ah al-Miṣrīyah al-ʿĀmmah lil-Kitāb: Bi-al-taʿāwun maʿa Markaz al-Buḥūth al-Amrīkī bi-Miṣr wa-Mu'assasat Smīthsūnyān, 1981).

24. For a traditional account of this role for Jesuits, see G. B. Nicolini, *History of the Jesuits: Progress, Doctrines, and Designs* (London, 1854), 96–133.

25. See, for example, the case of Manuel Godinho, a Portuguese traveler whose itinerary was similar to della Valle's, but with different motives. Manuel Godinho, *Intrepid Itinerant: Manuel Godinho and His Journey from India to Portugal in 1663*, ed. John Correia-Afonso, trans. Vitalio Lobo and John Correia-Afonso (Bombay: Oxford University Press, 1990).

26. Herbert Thomas, *Some Years Travels into Divers Parts of Africa and Asia the Great Describing More Particularly the Empires of Persia and Industan* (London, 1662), 40.

27. Ibid., 126, 129, 137.

28. Ibid., 238.

29. On the academy, see Laura Alemanno, "L'Accademia degli umoristi," *Roma moderna e contemporanea* 3, no. 1 (1995): 97–120.

30. Two biographical dictionaries of Calabrians mention Schipano as a physician and a poet: Luigi Aliquò Lenzi and Filippo Aliquò Taverriti, *Gli scrittori Calabresi: dizionario bio-bibliografico* (Reggio di Calabria Tip.: Corriere di Reggio, 1955), 3:205; Luigi Accattatis, *Le biografie degli uomini illustri delle Calabrie* (Cosenza, 1869), 2:409.

31. Bull, *Pilgrim*, xi.

32. Della Valle, *De viaggi*, 1:146–52. In the letter to Schipano, della Valle listed manifold Arabic books from jurisprudence through sciences, ending up with dictionaries, which he collected for Schipano and his circle in Naples.

33. Della Valle, *De viaggi*, 1:149–52.

34. Ibid., 1:150.

35. Ibid., 1:153–55.

36. Della Valle found the two portrait mummies at Saqqara. They dated to about the fourth century AD and showed a man and a woman. They were eventually sold by the estate of Count Chigi and acquired by the Municipal Office of Dresden in 1728, and they are now part of the Egyptian Section of Dresden's Art Gallery. See Bull, *Pilgrim*, xii; letter on Egypt in della Valle, *De viaggi*, 1:192.

37. Della Valle, *De viaggi*, 1:215–38; letter of March 7, 1616, in ibid., 1:239–50.

38. Della Valle, *De viaggi*, 1:309.

39. Ibid., 1:310.

40. Ibid., 1:328.

41. Ibid., 2:541

42. Della Valle, "Letter to al-Lārī," 2–3. Given that he mentions "the end of Kepler's life," it seems that the translation started in Goa in 1623 was finished in Rome after Kepler's death in 1630. This is further evidence that the version of this letter in the Vatican was completed in the early 1630s.

43. Ibid., 5.

44. Anant Kakba Priolkar, Gabriel Dellon, and Claudius Buchanan, *The Goa Inquisition: Being a Quatercentenary Commemoration Study of the Inquisition in India* (Bombay: Bombay University Press, 1961), 176.

45. Della Valle, "Letter to al-Lārī," 25.

46. This verse elicited medieval commentaries that were no doubt used by de Zuñiga. Rashi's twelfth-century exegesis reads: "God looks at earth and its shaking"; Metzudat David: "God moves earth from its place"; Biur Milim: "God tells the Sun not to rise or not to move"; and Biur 'Inyan: "God is the one that determines the generation and decay of earth, and shaking it means he would destroy it."

47. Translated by Victor Navarro Brotóns, in his "The Reception of Copernicus in Sixteenth-Century Spain," *Isis* 86, no. 1 (1995): 67. See also Richard Blackwell, *Galileo, Bellarmine, and the Bible* (Notre Dame, IN: Universiy of Notre Dame Press, 1991), 180–81.

48. Robert Westman, "The Copernicans and the Churches," in *God and Nature: Historical Essays on the Encounter between Christianity and Science*, ed. David Lindberg and Ronald Numbers (Berkeley and Los Angeles: University of California Press, 1984), 92.

49. On the principle of accommodation, see Amos Funkenstein, *Theology and Scientific Imagination: From the Middle Ages to the Seventeenth Century* (Princeton, NJ: Princeton University Press, 1986), 213–19; Carlo Ginzburg, *Wooden Eyes* (New York: Columbia University Press, 2001), 117–18, 145–47.

50. Agostino Giustiniani, *Psalterim Hebreum, Grecum, Arabicum, et Chaldeum cum tribus Latinis interpretationibus et glosis . . . Aug. Luistinaini genuensis praedictorii ordinis episcopi calendis Augustii* (Genoa, 1516).

51. Frank Rosenthal, "The Study of the Hebrew Bible in Sixteenth-Century Italy," *Studies in the Renaissance* 1 (1954): 81–89.

52. François Vavasseur, *Francisci Vauasseur e Societate Iesu. Iobus: Carmen Heroicum* (Paris, 1638).

53. Translation of de Zuñiga in Brotóns, "Reception of Copernicus," 64.

54. Borrus also cites the Book of Job in his *Collecta astronomica*, but he could not have been the source for della Valle. His interest was in other verses, such as Job 38:8–9, which were pertinent to his attempt to reconcile the biblical account of the creation with Aristotelian physics. Borrus writes: "Father Francisius Suarius [Francisco Suárez?] in *The Six Days of Creation* [*Opus sex dierum*] ch. 5, p. 8, relates that this same opinion was held by several authorities along with Pereira. These affirmed that the air was made on the second day in the following way: on the first day the whole space between the earth and the moon was filled not with the three other elements, but with a kind of cloudy material or thin vapor, intermediate as it were between Air and Water. They infer this not only from the aforementioned passage in Ben Sirach (a.k.a. Ecclesiasticus; Wisdom is speaking): 'As a cloud I covered all the earth,' but also from another one, in Job: 'Who shut up the sea with doors, when it broke forth as if issuing out of the womb: When I made a cloud the garment thereof, and wrapped it in a mist as in swaddling bands?' According to Suarius, they say that it was therefore from

this material that Air was made—and in fact Fire too—on the second day, through a kind of thinning and rarefaction caused by a divine miracle. Suarius quotes these statements from others, and though he himself tries to refute their opinion, nevertheless anyone who reads the arguments that he gives to the contrary will see that they have too little or no force." Borri, *Collecta astronomica*, 414–15. Borrus later repeats the same biblical quotations to make the same cosmological point (435–36).

55. Della Valle, "Letter to al-Lārī," 25–26.

56. See Yair Hoffman, *Shelemut pegumah: Sefer Iyov ve-rik'o* [Blemished Perfection: The Book of Job in Its Context] (Jerusalem: Biyalik Institute, 1995), 132.

57. Della Valle, "Letter to al-Lārī," 25–26.

58. He made this comment even though bibles in ancient Near Eastern languages were available in Europe. On the manuscripts and printed Chaldean-Aramaic versions existing in Europe in the fifteenth and sixteenth centuries, see David M. Stec, *The Text of the Targum of Job* (Leiden: Brill, 1994).

59. In his commentary on the Book of Job 2:12, Ibn-'Ezra also copes with the question of the fall of linguistic knowledge since Babel. Moreover, in his commentaries on different places in the Bible, he expresses his argument that there were original scriptures that preceded the Mesoraic tradition. See, for example, his commentary on Genesis 11:1 "the whole earth was one language," the introductory verse on the Tower of Babel. See also Ibn-'Ezra's commentary on Exodus 25:31, in which he notes that the word *tei'aseh* ("will be made") was "misspeld" with an additional letter *yod'*. For Ibn-'Ezra, the addition raised an important question about the source of Scripture; he writes: "I have seen books [i.e., manuscripts] which have been examined by the Masoretes of Tiberias, and fifteen of their elders have sworn that they have looked closely three times at each word and dot, and the word is written with a *yod* . . . however in manuscripts from Spain, France and overseas I did not find a *yod* . . . anyway if there is a *yod* there, it is a foreign word." For further reading, see Mariano Gómez Aranda, ed. and trans., *El comentario de Abraham Ezra al libro de Job* (Madrid: Consejo Superior de Investigaciones Científicas, Instituto de Filología, 2004).

60. In the sixteenth and seventeenth centuries, both the astronomical and exegetical writings of Ibn-'Ezra were translated and published in Latin. For instance, Archbishop Ussher relied on Ibn-'Ezra's biblical commentaries, especially on the Book of Job, to compose a biblical chronology. James Ussher, "Chronologia Sacra," in *The Whole Works of the Most Rev. James Ussher*, ed. Charles Richard Elrington (Dublin: Hodges, Smith, 1864), 12:50–51. For the availability of Ibn-'Ezra's commentary for Hebraist ideas, see Avraham Ibn-'Ezra, *In decalogum commentarius: Doctrina et eruditione non careens* (Paris, 1568); Avraham Ibn-'Ezra, *Hosee cum Thargum, id est Chaldaica paraphrasi Jonathan, et commentariis R. Abraham Ezra et R. D. Kimchi* (Geneva, 1556); and Peter Rooden, *Theology, Biblical Scholarship, and Rabbinical Studies in the Seventeenth Century* (Leiden: Brill, 1989), 9, 44. On the incorporation of the commentary of Ibn-'Ezra into Cabala, see Israel Ben Moses and Voisin de Joseph, *Dispvtatio cabalistica R. Israel filii R. Mosis de anima: et opus rhythmicum R. Abraham abben Ezræ de modis, quibus Hebræilegem solent interpretati verbum de verbo expressum extulit nobilis Ioseph de Voysin* (Paris, 1635). On Hebrew grammar

in Hebraism according to Ibn-'Ezra, see Pagnino Santi, *Institvtionvm hebraicarvm abbreuiatio* (Lyons, 1528). On Ibn-'Ezra's astrological writings, see Jean Ganivetus, *Epistola astrologie defensiva [Gondisalvi Toledo] Amicus medicorum magistri Johannis Ganiveti: cum opusculo quod Celi enarrant propter principium ejus inscribitur: et cum abbreviatione Abrahe Aveneezre de luminaribus et diebus creticis. Astrologia Ypocratis* (In civitate Lugdunense, 1508); and Jean Ganivetus, *Opusculum repertorii prognosticon in mutationes aeris* (1485). By the early eighteenth century, his full commentary on the Bible was also printed in Latin in Lundius Daniel, *Commentarius R. Aben Esrae in Prophetam Habacuc: Quem ex Hebraeo in Latinum sermonem versum & brevibus notis illustratum* (Upsal, 1706).

61. On this tradition, see N. H. Tur Sinai, *The Book of Job: A New Commentary* (Jerusalem: Kiryat Sefer, 1967), xxxi, 111. See also Hoffman, *Shelemut pegumah*, 183–223.

62. Della Valle, "Letter to al-Lārī." The "word" that della Valle mentions is *shaken*; the Hebrew is מרגיז and the Chaldean or Aramaic is דמרגיז, both usually translated as "upsetting"; some medieval commentators—for example, Rashi—suggest other meanings such as "shaking" or "moving."

63. See *Catholic Encyclopedia* (New York: Robert Appleton, 1908), s.v. "Chaldean Christians." See also Charles J. Borges, *The Economics of the Goa Jesuits, 1524–1759: An Explanation of Their Rise and Fall* (New Delhi: Concept Publishing, 1994).

64. On Galileo and de Zuñiga, see Francesco Barone, "Deigo De Zuñiga e Gelileo Galilei: Astronomia eliostatica ed esegesi biblica," *Critica storia* 3 (1982): 319–34.

65. From Blackwell, *Galileo, Bellarmine, and the Bible*, 122.

66. Paolo Antonio Foscarini, *Concerning the Pythagorian and Copernican Opinion of the Mobility of the Earth and Stability of the Sun and of the New Systeme or Constitution of the World*, in *Mathematical Collections and Translations . . .*, ed. Thomas Salusbury (London, 1667), 475.

67. It is noteworthy that Naples of the early sixteenth century had a lively Jewish and Hebrew culture. Some of the first Hebrew printing presses of Italy were in Naples, and the first printing of Avicenna's *Canon of Medicine* was made in 1491 by the Ashkenazi press. However, this printing culture was shut down by the middle of the sixteenth century when the Jews of Naples were expelled. On Hebrew printers in Naples, see David Werner Amram, *The Makers of Hebrew Books in Italy* (London: Holland Press, 1963), 63–69.

68. Jerome Friedman, *The Most Ancient Testimony: Sixteenth-Century Christian-Hebraica in the Age of Renaissance Nostalgia* (Athens: Ohio University Press, 1983), 72–75.

69. For a survey of sources and documents on Foscarini's life and works, see Emanuele Boaga, "Annotazioni e documenti: sulla vita e sulle opera di Paolo Antonio Foscarini teologo 'Copernicano (1562c.–1616),'" *Carmelus* 37 (1990): 173–216. One of the few clues about Foscarini's motives to write his work concerns another enigmatic Neapolitan, Vincenzo Carafa. According to Foscarini, Carafa ordered him to write and publish the great compromise of Scriptures with Copernicanism for the defense of Galileo. Some scholars identify Carafa as the famous Jesuit professor at the College

of Naples who was eventually appointed as seventh general of the Society of Jesus (1646–49). See Stefano Caroti, "Un sostenitore napoletano della mobilità della terra: il padre Paolo Antonio Foscarini," in *Galileo e Napoli,* ed. Fabrizio Lomonaco and Maurizio Torrini (Naples: Università degli Studi di Napoli, 1987), 97; Boaga, "Annotazioni," 183–94. Indications supplied by Foscarini, however, suggest the existence of another Vincenzo Carafa, son of Fabrizio, quarter count of Ruvo, and a knight of the St. John Jerusalem Order. But by 1615 (the year of publication of Foscarini's treatise), this Vincenzo Carafa had been dead for four years. See B. Aldimari, *Historia genealogica della familiglia Carafa* (Naples: Bulifon, 1691), 3:83–90. It is uncertain that this Carafa was Carafa the Jesuit, because the Vincenzo Carafa who became seventh general of the Society of Jesus did not belong to the St. John Jerusalem Order. This confusion may simply indicate the Neopolitan practice whereby men of the apostolic missions appealed to Neapolitan nobility by sometimes carrying cavalry titles. See D. Bartoli, *Della vita del P. Vincenzo Carafa, settimo generale della compagnia di Giesú* (Rome, 1651). Although the *Catholic Encyclopedia* mentions certain writings of Carafa, the Society of Jesus general, it does not mention that he encouraged Foscarini to write the compromise. Thus, if we deduce that the Carafa of interest to us here was *not* the seventh general of the Jesuits, this might explain the lack of biographical evidence on him—a result of his membership and practices in secret circles. *Catholic Encyclopedia,* s.v. "Caraffa, Vincent."

70. Galileo made this statement in a letter to the grand duke through the Tuscan secretary of state, Curzio Picchena, who had conveyed the news on the decision of the Inquisition. Galileo, *Opere,* ed. Saragat Giuseppe (Florence: G. Barbèra, 1968), 12:243–45.

71. Tommaso Campanella, *Apologia per Galileo: testo Latino a fronte, acura di Paolo Ponzio* (Milan: Rusconi, 1997),172–73. Campanella mainly cited Ambrose, according to whom Pythagoras taught that we should abstain from certain foods; that there is only one God, even though he said that the angels are secondary gods; and that all things are numbers (as did Moses in the construction of the Tabernacle, and as did Solomon, who said [Wisdom 11:21] that all things were created "in numbers, weight, and measure"). For Christian fathers, such as Ambrose, Pythagoras was an authoritative figure because he was thought to have been a Jewish intermediary between Moses and the Christian community. See St. Ambrose, Bishop of Milan, *Epistolae et opuscula* (Milan: Antonio Zarotto, 1491). The belief in the Jewishness of Pythagoras had been spread since the first century, particularly by the Roman Jewish historian Josephus Flavius, who stressed the Jewish origins of Pythagoras in his book on Jewish antiquities, accentuating that even Greek philosophy emanated from ancient Judaism. Josephus Flavius, *Antiquitates Judaicae,* trans. Ralph Marcus (Cambridge, MA: Harvard University Press, 1998). There are several English translations of Campanella's apologia. See, for example, Tommaso Campanella, *Apologiae pro Galileo: A Defense of Galileo the Mathematician from Florence, Which Is an Inquiry as to Whether the Philosophical View Advocated by Galileo Is in Agreement with, or Is Opposed to, the Sacred Scriptures,* trans. Richard J. Blackwell (Notre Dame, IN: University of Notre Dame Press, 1994); Tommaso Campanella, "The Defense of Galileo," trans. Grant

McColley, *Smith College Studies in History* 22, no. 3–4 (1937). The view that Pythagoras was Jewish was shared by other Copernicans of the early seventeen century. John Wilkins, who wrote a book about the habitable moon, asserted that "some think him [Pythagoras] a *Iew* by birth, but most agree that hee was much conversant amongst the learneder sorr, & Priests of that Nation, by whome he was informed of many secretes." John Wilkins, *The Discovery of a World in the Moone; Or, A Discourse Tending to Prove That 'Tis Probable There May Be Another Habitable World in That Planet* (London: Printed by E.G. for Michael Sparke and Edward Forrest, 1638), 81.

72. Campanella, *Apologia per Galileo*, 60–61, 172–73.

73. Ibid., 174–75.

74. For more details on this fascinating story, see Noel Malcolm, "The Crescent and the City of the Sun: Islam and the Renaissance Utopia of Tommaso Campanella," *Proceedings of the British Academy* 125 (2005): 41–67.

75. See Biagioli, *Galileo, Courtier*, 32; Galileo, *Opere*, vol. 10.

76. In accounts of Neapolitan scholarship, Schipano is described as a doctor and a friend of della Valle. See Giangiuseppe Origlia Paolino, *Istoria dello studio di Napoli* (Naples, 1753), 2:415. In a general history of Naples, Schipano is mentioned as a doctor interested in botany and, once again, as a friend of della Valle. Pietro Giannone, *The Civil History of the Kingdom of Naples*, trans. James Ogilvie (London, 1723), 2:715, 843.

77. Schipano's guest book entry in MS 30, p. 72v, Lincei Archive, Rome.

78. Giuseppe Olmi, "La Colonia Lincea di Napoli," in Lomonaco and Torrini, *Galileo e Napoli*, 50.

79. Ibid., 54.

80. On the connection between the members of the Lincei Society and the Galilean debate, see Giovanna Baroncelli, "L'astronomia a Napoli al tempo di Galileo," in Lomonaco and Torrini, *Galileo e Napoli*, 197–225. On the connection of Colonna to Galilean discoveries, see ibid., 205–10.

81. Baroncelli, "L'astronomia," 209. Moreover, Colonna's interests necessitated a knowledge of Islamic languages. Diego de Urrea Conca, another Neapolitan and an Arabist, joined the Lincei Society five days after Colonna, on January 27, 1612. The Linceans needed an expert Arabist, at least partly to translate classical works of Arabic science. They enrolled Diego de Urrea Conca, who, as interpreter at the court of Fez and for the king of Spain, knew Arabic, Turkish, and Persian. See David Freedberg, *The Eye of the Lynx: Galileo, His Friends, and the Beginnings of Modern Natural History* (Chicago: University of Chicago Press, 2002), 114. See also Giuseppe Gabrieli, *Contributi alla storia della Accademia dei Lincei* (Rome: Accademia nazionale dei Lincei, 1989); Giuseppe Gabrieli, *I primi Accademici Lincei e gli studi orientali* (Florence: Olschki, 1926).

82. Gabriella Belloni Speciale, "La Ricerca botanica dei Lincei a Napoli," in Lomonaco and Torrini, *Galileo e Napoli*, 60, 74, 75, 77.

83. Baroncelli, "L'astronomia," 200.

84. Giannone, *Civil History*, 2:715. For a description of the 1616 reforms at Naples University, see ibid., 712–15.

85. See Blackwell's introduction to Tommaso Campanella in *Apologiae pro Galileo*, 9. In addition, one of the few sources we have on Schipano is an entry under his name in a biographical dictionary of the Lincei Society, where he is called a friend of Campanella. Gabrieli, *Contributi alla storia dell'Accademia dei Lincei*, 2:1523–29.

86. Luigi Amabile, *Fra Tommaso Campanella, la sua congiura, i suoi processi e la sua pazzia* (Naples: A. Morano, 1882), 1:7.

87. Luigi Amabile, *Fra Tommaso Campanella ne' castelli di Napoli, in Roma e in Parigi* (Naples: A. Morano, 1887), 1:171. For other connections between Schipano and Campanella, see ibid, 1:93, 151, 171; and 2:38, 133.

88. Bellori, *Vita di Pietro della Valle il pellegrino*, 9–10.

Three · Transcending Time in the Scribal East

1. Joseph Solomon Delmedigo, *Sefer Elim* (Amsterdam, 1628 [reprint, Odessa, 1865]).

2. See Geiger's introduction in Joseph Solomon Delmedigo, *Melo hofanim*, ed. Abraham Geiger (Berlin, 1860).

3. Isaac Barzilay, *Yoseph Shlomo Delmedigo (Yashar of Candia)* (Leiden: Brill, 1974), 48.

4. Joseph Levi, "Haakdemya hayehudit lemada bitehilat hameah hasheva-ʿesre: hanisayyon shel Shlomo Delmedigo" [Jewish Academy for Sciences in the Early Seventeenth Century: The Experience of Solomon Delmedigo], in *The Eleventh International Congress for Jewish Studies*, Section B (Jerusalem, 1984).

5. Frances Amelia Yates, *Giordano Bruno and the Hermetic Tradition* (Chicago: Chicago University Press, 1964); David Ruderman, *Jewish Thought and Scientific Discovery in Early Modern Europe* (New Haven, CT: Yale University Press, 1995), 118–53. Ruderman cites unpublished papers by Joseph Levi that suggest Levi's view of Delmedigo's works encompasses aspects of natural philosophy, the Cabala, magic, and the critique of Jewish intellectual culture (134).

6. Joseph Kaplan, "'Karaites' in Early Eighteenth-Century Amsterdam," in *Sceptics, Millenarians, and Jews*, ed. David Katz and Jonathan Israel (Leiden: Brill, 1990), 196–237.

7. Richard Popkin, "The Lost Tribes, the Caraites and the English Millenarians," *Journal of Jewish Studies* 37, no. 2 (1986): 213–27. See also Richard Popkin, "Les Caraites et l'emancipation des Juifs," *Dix-Huitième Siècle* 13 (1981): 137–47.

8. Joseph Solomon Delmedigo, *Maamar ʿal kochav shavit* [An Article on a Comet], F 64619, siman 0325/3, p. 1, Institute of Hebrew Manuscripts, Hebrew National Library, Jerusalem.

9. Ibid.

10. Ibid., 6.

11. Joseph Solomon Delmedigo, "Gevurot haShem," in *Elim*, 301. Page numbers for *Elim* here and throughout refer to the 1865 reprint.

12. Joseph Solomon Delmedigo, "Mayan hatum," in *Elim*, 417. He says that "we," the students of Galileo, "used to look through the Sheforfert of glass" (434).

13. Delmedigo's student Moshe Metz, who wrote an introduction to *Elim*, tells us that "in Cairo he [Delmedigo] was associated with Jacob Iskandrani, who is also a great scholar, and compiled a book of the wonders of god in the art of mechanics. He did not reach to the highest level of wisdom because of different opinions in religion and belief [being a Karaite], but it did not disturb him [Delmedigo] to be associated with any scholar, whoever he was, that was interested in reason." Metz, introduction to Delmedigo, *Elim*, ix.

14. Delmedigo, "Gevurot haShem," 320.

15. Delmedigo, *Melo hofanim*, 13.

16. Delmedigo, "Mayan Hatum," 376.

17. Delmedigo, *Elim*, xiv.

18. We know only a few details about 'Ali Ben Raḥim al-Dīn from Delmedigo's account. I found no other information in the biographical dictionary by Kaḥḥāla 'Umar Riḍa, *Mu'jam al-mu'allifīn: tarājim muṣannifī al-kuttub al-'Arabiyyah* (Beirut: Muassasat al-Risalah, 1993), or in David King, ed., *Fihris al-makhṭāṭāt al-'ilmīyah al-maḥfāẓah bi-Dār al-Kutub al-Miṣrīyah* [Catalogue of the Scientific Manuscripts in the Egyptian National Library] (Cairo: al-Hay'ah al-Miṣrīyah al-'Āmmah lil-Kitāb: Bi-al-ta'āwun ma'a Markaz al-Buḥāth al-Amrīkī bi-Miṣr wa-Mu'assasat Smīthsānyān, 1981).

19. All extracts here are from Joseph Solomon Delmedigo, "Sod haYasod," in *Elim*, 176–77.

20. In Delmedigo's self-reference as "boy" (as quoted in the text, from his "Article on a Comet"), even though he was in his late twenties, he fashioned a persona by invoking two common myths embedded in his name. First was the biblical Joseph, a captive boy who arrived in Egypt and paved a way to become second in power to the king. Second was the biblical Solomon, a wise king who struggled to integrate the kingdom into its surroundings. The resemblance between the myths and Delmedigo's rhetoric suggests that, like Joseph, he was "a boy" who went to Egypt and struggled to the top of the intellectual culture and saved the Jews from further humiliation; and, like Solomon the Wise, he was, as "a boy," the most important Jewish natural philosopher. We learn more about this in the first pages of *Elim*, which includes a short poem that plays with the names "Solomon" and "Joseph" and encapsulates Delmedigo's self-image.

21. Metz, introduction to Delmedigo, *Elim*, viii, ix.

22. Ibid., xv.

23. Ibid., x.

24. Ashkenazi (Delmedigo's student), introduction to Shlomo Yosef [Joseph Solomon] Delmedigo, *Novlot ḥokhmah* (1631) (Tel Aviv, 1970).

25. See M. A. Shulvass, "Sfarim vesifriot bekerv yehude Italia barenesans" [Books and Libraries among Italian Jews during the Renaissance], *Talpiot* 4 (1910): 591–605; I. Sonne, "Book Lists through Three Centuries: A First Half of the Fifteenth Century, Italy," *Studies in Bibliography and Booklore* 2 (1955): 3–19; A. M. Habermann, *Ha-Madpiss Cornelio Adel Kind u-beno Daniel: u-reshimath haSefarim sheNidpessu al yedehem* [The Printer Cornelio Adel Kind, His Son Daniel and a List of Books Printed

by Them] (Jerusalem: Rubin Mass, 1980); Yzhak Yudlov, ed., *Giouvanni Di Gara: Printer, Venice 1564–1610, List of Books Printed at His Press* (Jerusalem: B. Habermann, 1982); Abraham Meir Habermann, *Ha-Madpis Daniyel Bombirg u-reshimat sifre bet defuso* [The Printer Daniel Bomberg and the List of Books Published by His Press] (Tsefat: Ha-Muzeon le-omanut ha-defus, 1978). Robert Bonfil suggests that the lack of printed works of philosophy indicates the decline of interest in philosophy within Jewish intellectual culture. Robert Bonfil, *Rabbis and Jewish Communities in Renaissance Italy* (London: Littman Library of Jewish Civilization, 1993), 275.

26. Delmedigo, *Elim*, 129.

27. Ibid., viii.

28. Metz tells us that Delmedigo lacked the incentive to visit the Holy Land because "from his early stages as a scholar, the wisdom of the Cabala was strange to him, and when he was close to the Holy Land, where the Cabala wisdom grew, shaping secrets, surely his soul strived to beg the ashes of Jerusalem, but it could not be fulfilled." Metz, introduction to Delmedigo, *Elim*, ix.

29. Delmedigo, *Melo hofanim*, 23.

30. One of the few works in Hebrew that mentioned Tycho Brahe was a manuscript written by a sixteenth-century Karaite, Yoseph Tishbi of Constantinople. His work, basically an account of observations and calculations of the lunar calendar, mentions the comet of 1577 and the revolutionary interpretation of Tycho. Yoseph Tishbi, *Hezionot vezihronot* [Visions and Memoirs] (Constantinople, 1579), F 21357, Institute of Hebrew Manuscripts, Hebrew National Library, Jerusalem.

31. Nathan Schur, *The Karaite Encyclopedia* (Berlin: Peter Lang, 1995), 210.

32. *Al-Muhtawi*, by the Karaite thinker Yusuf al-Basri (d. 1040), stresses the common denominator between Karaism and Mu'tazilite rationalism and underlines the rational character of ethics and the priority of reason over tradition. Al-Basri confirmed that man has a free will, but God knows what his choice will be. Yusuf al-Basri, *Al-Kitab al-Muhtawi de Yusuf Al-Basir* (Leiden: Brill, 1985), 149, 255, 311, 316, 344, 439, 609. Another scholar associated with Mu'tazilite philosophy and atomism was Jeshua ben-Jehudah, who held that knowledge of the creation cannot be derived from Scripture alone, but is subject also to rational speculation. See Leon Nemoy, *Karaite Anthology* (New Haven, CT: Yale University Press, 1952), 123–32.

33. Alnoor Dhanani, *The Physical Theory of Kalam: Atoms, Space, and Void in Basrian Mutazili Cosmology* (Leiden: Brill, 1994), 54–89; Alnoor Dhanani, "Kalam and Hellenistic Cosmology: Minimal Parts in Basrian Mu'tazili Atomism" (Ph.D. diss., Harvard University, 1991).

34. S. A. Poznanski, *The Karaite Literary Opponents of Saadiah Gaon* (London: Luzac, 1908), 46–48. See also Schur, *Karaite Encyclopedia*, 47, 210.

35. Jeshua ben-Jehudah was called, in Arabic, Abu al-Faraj Farkhan ibn Asad. See Nemoy, *Karaite Anthology*, 123–32; Schur, *Karaite Encyclopedia*, 161–62.

36. Daniel J. Lasker, "Aaron Ben Joseph and the Transformation of Karaite Thought," in *Torah and Wisdom: Studies in Jewish Philosophy, Kabbalah, and Halacha: Essays in Honor of Arthur Hyman*, ed. Ruth Link-Salinger (New York: Shengold, 1992), 121–28.

37. Aaron ben Elijah, *Etz hayyim* (Leipzig: J. A. Barth, 1841); Daniel Lasker, "Nature and Science according to Aaron ben Elijah the Karaite," *Da'at* 17 (1986): 33–42; Schur, *Karaite Encyclopedia*, 11.

38. Delmedigo, *Melo hofanim*, 19.

39. Joseph Solomon Delmedigo, *Mitzraf ḥokhmah* (Warsaw: ha-Madpis D. Ṭorsh, 1890), 51; Delmedigo, "Gevurot haShem," 343.

40. Delmedigo followed the tradition that Ibn-'Ezra relied on Karaite commentary, a tradition that was stressed by Comtiano. See Jean-Christophe Attias, *Le Commentaire biblique: Mordechai Komtiano ou l'heméneutique du dialogue* (Paris: Cerf, 1991). However, later Karaite commentators relied on Ibn-'Ezra for parallels with the Karaites. See Frank Daniel, "Abraham Ibn 'Ezra and the Karaite Exegetes Aharon Ben Joseph and Aharon Ben Elijah," in *Abraham Ibn Ezra and His Age*, ed. Fernando Días Esteban (Madrid: Asociación Española De Orientalistas, 1990), 99–109.

41. Pico della Mirandola, who played a large role in the life of a Delmedigo intellectual ancestor, Tommaso Campanella (see chapter 2), had in his library the astrological compilations of Ibn-'Ezra. See Pearl Kibre, *The Library of Pico della Mirandola* (New York, Columbia University Press, 1936), items 879, 1016, 1055.

42. Robert Westman, "Two Cultures or One? A Second Look at Kuhn's *The Copernican Revolution*," *Isis* 85, no. 1 (1994): 89; Avraham Ibn-'Ezra, *Liber de revolutionibus et nativitatibus* (Venice, 1507).

43. A collection of Ibn-'Ezra's astronomical essays was published in Latin in 1507. Avraham Ibn-'Ezra, *Opera* (Venice, 1507).

44. See Yosef Cohen, *Haguto hafilosofit shel Rabbi Avraham Ibn-'Ezra* [The Philosophical Thought of Avraham Ibn-'Ezra] (Ramleh: Hish, 1996), 347–45.

45. Delmedigo also read a manuscript (perhaps obtained in Istanbul) by Moshe Almosnino that was a translation of Sacrobosco's *Beit elim* (in Hebrew). He also consulted Almosnino's *Shaar HaShammyyim*, a translation of Peurbach. Joseph Solomon Delmedigo, "Hukot shamayyim," in *Elim*, 244. See also Delmedigo, "Gevurot haShem," 306; Delmedigo, "Mayan hatum," 369, 429. Almosnino is also mentioned in Delmedigo, *Melo hofanim*, 13.

46. A manuscript by one of Istanbul's prominent astronomers, 'Alī Qūshjī, was available in Hebrew. A manuscript by 'Alī Qūshjī in Judeo-Persian (Persian in Hebrew letters) is in the British Library Add. 7701, pp. 38–82, catalogued as no. 12365 in the Institute of Hebrew Manuscripts, Hebrew National Library, Jerusalem.

47. Delmedigo mentions in *Elim* the different translations of Ptolemy, Averroes, al-Battani, and others; some were filled with mistakes and "we did not know to whom should be ascribed the mistakes, whether to the stubbornness of our understanding or [to] the work of the sloppy copiers." Moreover, he tells us that, in Istanbul, he worked together with other scholars on the different accounts of Ptolemy's *Almagest*, but when they reached the fourteenth chapter they could not understand the arguments and therefore referred, instead, "to the work of Jabir Ben Alfak al-Shabily who facilitated our understanding using the art of disentangling triangles." This indicates several things: first, in Istanbul, Delmedigo was engaged in astronomical research with local astronomers; second, these scholars referred to Arabic works to facilitate

their understanding of Ptolemy; third, Delmedigo made critical remarks about the problems of manuscripts; and fourth, he searched in Egypt and Istanbul for early sources of astronomy in Arabic that might prevent mistakes in translation. Delmedigo, *Elim*, 19.

48. Delmedigo writes: "Many Rabbis these days would not understand things regarding the spheres and the like. And they would not look at the deeds of God and the creation. They never researched. [But they should remember that] as it is written, 'raise your heads to heavens' and see who created these." Delmedigo, *Melo hofanim*, 13.

49. Delmedigo, *Elim*, xiv.

50. Ibid. We do not know whether Metz or Delmedigo wrote this part of the introduction.

51. Ibid., 62.

52. Introduction to Delmedigo, "Sod haYasod," 130–31.

53. *Mutus liber, in quo tamen tota philosophia hermetica, figuris hieroglyphicis depingitur* (La Rochelle: P. Savouret, 1677).The *Mutus liber* stimulated much speculation about its meaning and purpose. Some scholars created an aura of impenetrable obscurity around the book and, as such, held it up as a sacrament, without understanding it. However, *Mutus liber* is important not just in its lack of writing, but in its emphasis on nonliterate symbols and meanings. Its hieratic symbols reserved for the elect the mysteries of science, with the *mutae artes*—that is, the silent or symbolic arts—meant for the illiterate. *Mutus liber*, therefore, is a book like all others and can be plainly read. All alchemical works, in verse or prose, in Latin or any other language, employed esoteric writing. One form of esoteric writing is to use letters of the common alphabet and vocabulary that must be decoded. Another form involves illustrations. The illustrated method, as used in *Mutus liber*, is the more transparent, given that the objective image is certainly more expressive than literary tropes and rhetorical figures, especially in a matter as experimental as chemistry.

54. Delmedigo, *Elim*, xii.

55. See commentaries by Rashi and Ramban on this verse.

56. See commentary by Ibn Saula 'Or Ganuz' in Nehunya Ben ha-Kanah, *Sefer haBahir* (Jerusalem: Bakal 1974), 79.

57. Moses Cordovero, *Sefer Elima* (Jerusalem: Or Hadash, 1999).

58. For Scholem's marginalia, see a special edition of the Zohar: Gershom Gerhard Scholem, *Sefer haZohar shel Gershom Scholem 'im he 'arot bi-khetivat yado* [The Zohar Book of Gershom Scholem with Notes in His Handwriting] (Jerusalem: Hebrew University Press, 1982), 3:1323. Scholem refers to *Midrash Raba Bamidbar* 19:3. There are also other cabalistic traditions that might have operated in Delmedigo's esoteric writing. For instance, we learn that after darkness, the sacred word *elohim* (meaning "god") may be thought of as losing, because of the diminution in light, three of its letters, thus resulting in the word *elim*. This was the cabalistic explanation for the tradition that Jews do not study sacred scholarship after sunset, a time of day reserved for secular and scientific learning.

59. The number of *elim* is actually 81.

60. Ben Natan's letter in Delmedigo, *Elim*, 25.

61. Delmedigo, *Elim*, 119.

62. Delmedigo, "Gevurot haShem," 299.

63. Ibid.

64. Delmedigo, *Elim*, 58.

65. Metz, in Delmedigo, *Elim*, 61–62.

66. Ibid., 61–62.

67. Ibid., 62.

68. Delmedigo, "Sod haYasod," 151.

69. See the introduction to Delmedigo, *Elim*.

70. Delmedigo, "Gevurot haShem," 348.

71. See, for example, John Collet, *A Treatise of the Future Restoration of the Jews and Israelites to Their Land: With Some Account of the Goodness of the Country, and Their Happy Condition There, Till They Shall Be Invaded by the Turks: With Their Deliverance from All Their Enemies, When the Messiah Will Establish His Kingdom at Jerusalem, and Bring in the Last Glorious Ages* (London: Printed for J. Highmore, M. Cooper, and G. Freer, 1774).

72. The book was *Seder haTefilot keMinahag haKaraim*, the prologue of which mentions that it was printed by the affluent gentleman Yosef Moshe Rabizi. See Habermann, *Ha-Madpiss Cornelio Adel Kind*, 66, item 145.

73. See Marion Leathers Kuntz, "Venezia portava el fuocho in seno: Guillaume Postel before the Council of Ten in 1548; Priest Turned Prophet," *Studi Veneziani*, n.s., 33 (1997): 96–97.

74. See Marion Leathers Kuntz, *Guillaume Postel: Prophet of the Restitution of All Things; His Life and Thought* (The Hague: Martinus Nijhoff, 1981), 96.

75. Ibid., 112, 118, 132.

76. George Saliba examined Arabic manuscripts in the Vatican Library and the Bibliotheque Nationale that were purchased by Postel in Istanbul. George Saliba, "Writing the History of Arabic Astronomy: Problems and Differing Perspectives," *Journal of the American Oriental Society* 116, no. 4 (1996): 709–19.

77. Popkin, "Lost Tribes"; Popkin, "Les Caraites et l'emancipation des Juifs."

78. John Dury, "An Epistolicall Discours, to Mr. Thorowgood, Concerning his Conjecture That the Americans Are Decended from the Israelites," in *Jews in America, or Probabilities That the Americans Are of That Race*, ed. Thomas Thorowgood (London, 1650), e2. See also Popkin, "Lost Tribes," 215.

79. Menasseh Ben-Israel, *The Hope of Israel: The English Translation of Moses Wall, 1652; Edited, with Introduction and Notes by Henry Méchoulan and Gérard Nahon; Introduction and Notes Translated from the French by Richenda George* (1650) (Oxford: Published for the Littman Library by Oxford University Press, 1987).

80. Durry, "Epistolicall Discours," e4.

81. Rittangel translated, from Hebrew to Latin, a central cabalistic work entitled *Book of the Creation* [*sefer HaYetzirah*], which is ascribed to Avraham the Patriarch. Johann Stephan Rittangel, *Sefer Yetsirah; id est, liber Iezirah qui Abrahamo Patriarchæ*

adscribitur, unà cum commentario rabi Abraham f. D super 32 semitis sapientiæ, à quibus liber Iezirah incipit (Amsterdam: apud Ioannem & Iodocum Ianssonios, 1642).

82. See Pierre Bayle, *Dictionnaire historique et critique* (London: Printed for J. J. and P. Knapton . . . , 1734–38), 4:880–81. See also Popkin, "Lost Tribes," 219.

83. Kaplan, "Karaites."

84. Ibid., 203–4.

85. Leon da Modena (Judah Arieh of Modena), *Sheelot uTeshvot zikney Yehuda*, ed. Shlomo Simpson (Jerusalem, 1957), letter no. 77, p. 99. See also Kaplan, "Karaites," 205.

86. Leon da Modena, *Letters of Rabbi Leon Modena*, ed. Yacob Boksenboim (Tel Aviv: Tel Aviv University Press, 1984), 337.

87. See Simonson's introduction in Shlomo Simonson, ed., *Sheelot u-teshuvot Zikne Yehudah* (Jerusalem: Mosad ha-Rav Kuk, 1955), 16, n. 37.

88. Da Modena, *Sheelot uTeshvot zikney Yehuda*, letter no. 77, p. 99.

89. See Cecil Roth, "Leone da Modena and England," *Transactions of the Jewish Historical Society of England* 11 (1929): 206–27; Cecil Roth, "Leone da Modena and the English Correspondents," *Transactions of the Jewish Historical Society of England* 17 (1953): 39–43.

90. Da Modena makes the association between Karaites and Protestants in *The History of the Present Jews throughout the World: Being an Ample tho Succinct Account of their Customs, Ceremonies, and Manner of Living, at this Time. Translated from the Italian, Written by Leo Modena, a Venetian Rabbi. To Which Are Subjoin'd Two Supplements, One concerning the Samaritans, the Other of the Sect of the Carraites, from the French of Father Simon, with His Explanatory Notes. By Simon Ockley, Vicar of Swavesey in Cambridgeshire* (London: E. Powell, 1707), 243.

91. Da Modena thought that critics of Judaism led Christians to have an interest in Karaites. He writes: "those criticks of ours who have endeavoured to give an account of the Carraites, upon the credit of the writings of the Jews have fallen into great mistakes." Da Modena, *History of the Present Jews*, 245. See also Cecil Roth, "Leon da Modena and the Christian Hebraists of His Age," in *Israel Abrahams Memorial Volume* (Vienna: Adolf Holzhausen's Successors, 1927), 384–401.

92. Jacques Gaffarel, *Unheard-of Curiosities: Concerning the Talismanical Sculpture of the Persians; the Horoscope of the Patriarkes; and the Reading of the Stars*, trans. Edmund Chilmead (London: G.D. for H. Moseley, 1650). See also the entry for Jacques Gaffarel in *Biographie universelle* (Paris: Delagrave, 1870–73), 347–48.

93. In the 1650 edition (the first in English), Gaffarel's introduction explains how the printed work of da Modena would help disenchant Christians of their opinion of Judaism and would show the usefulness of certain Jewish rituals. Gaffarel's introduction to Leon da Modena, *The History of the Rites, Customes, and Manner of Life, of the Present Jews, throughout the World. Written in Italian, by Leo Modena . . . Tr. into English, by Edmund Chilmead* (London, 1650), viii–xxvi.

94. Da Modena argues that Christians who were pro-Karaite were generally so for polemic purposes; he differentiates Morin from other Christians, because Morin was

consulting a Karaite text of Aaron Ben-Joseph (Ahron Ben-Josef): "I know not from whence a doctor of the Sorbonne, their King's professor of Hebrew, has taken what he has written not long since upon this subject concerning the carrites." Da Modena, *History of the Present Jews*, 246.

95. On Jean Morin, see *Biographie universelle*, 29:327–29. See also Morin's *Commentarius Historicus de Disciplina in Administratione Sacramenti Poenitentiae Tredecim Primis Seculis in Ecclesia Occidentali, et huc usque in Orientali Obseruata, in Decem Libros Distinctus* (Paris: G. Meturas, 1651).

96. Da Modena, *History of the Present Jews.*

97. See Mark Cohan, trans. and ed., *Autobiography of a Seventeenth-Century Rabbi: Leon Modena's Life of Judah* (Princeton, NJ: Princeton University Press, 1988), 6.

98. In a 1611 letter to a friend, da Modena writes: "I would like to reveal a secret. I have here with me a brave man who came from a far and distant land, who showed me alchemically made silver." Boksenboim, *Letters of Rabbi Leon Modena*, 98.

99. Cohan, *Autobiography of a Seventeenth-Century Rabbi*, 111–14.

100. Da Modena did not publish *Voice of the Fool* until 1622, and it was ascribed to a Yedaiah ibn Raz (the knower of God and the son of secret). I used the text of Kol Sekhal included in Leon da Modena, *Behinat haCabala* (Jerusalem: Sifriyah le-maḥshevet Yiśra'el, 1968).

101. Ibid., 5.

102. Ibid., 21.

103. Ibid., 47, 63.

104. See da Modena's approbation in Delmedigo, *Elim*, 4.

105. Delmedigo, *Elim*, xiv.

106. Ibid., 135.

107. Simone Luzzatto, *Socrate overo dell'humano sapere* (Venice: Tomasini, 1651).

108. Simone Luzzatto, *Discorso circa il stato degli Hebrei* (Venice: Appresso Gioanne Calleoni, 1638). I used the Hebrew translation, Luzzatto, *Maamar 'al yehudei Venetzia*, trans. Dan Lattas (Jerusalem: Bialik Institute, 1950). See also Benjamin Ravid, *Economics and Toleration in Seventeenth Century Venice: The Background and Context of* Discorso *of Simone Luzzato* (Jerusalem: American Academy for Jewish Research, 1978).

109. Luzzatto, *Maamar 'al yehudei Venetzia*, 138.

110. Ibid.

111. Ibid., 142.

112. Luzzatto's approbation in Delmedigo, *Elim*, 4.

113. Luzzatto, *Maamar 'al yehudei Venetzia*, 148.

114. Ibid.

115. HaLevi's approbation in Delmedigo, *Elim*, 5.

116. Ben-Israel, in Delmedigo, *Elim*, 125. An example that could confirm Ben-Israel's complaints is the mistake in the year when a comet appeared: 1577 was mistyped as 1557. Throughout the book there are mistakes and misspellings of names, such as al-Amnun instead of al-Mamun. See Delmedigo, "Gevurot HaShem," 334.

117. See J. D. Ancona, *Delmedigo, Menasseh ben Israel en Spinoza* (Amsterdam: Hadapas, n.d.).

Four · Converting Measurements and Invoking the "Linguistic Leviathan"

1. Steven Shapin and Simon Schaffer, *Leviathan and the Air Pump: Hobbes, Boyle, and the Experimental Life* (Princeton, NJ: Princeton University Press, 1985).

2. For example, Mercier suggests that Greaves's project might have legitimized patrons and rulers. Raymond Mercier, "English Orientalists and Mathematical Astronomy," in *The "Arabick" Interest of the Natural Philosophers in Seventeenth-Century England*, ed. G. A. Russel (Leiden: Brill, 1994), 158–77. In the same vein, but from an institutional viewpoint, Feingold suggests that the motives inherent in combining natural philosophy and Orientalism were "to set the intellectual life of Oxford on a new footing" and to reformulate the nature and character of both undergraduate and graduate learning. Mordechai Feingold, "Patrons and Professors: The Origins and Motives for the Endowment of University Chairs—in Particular the Laudian Professorship of Arabic," in Russel, *"Arabick" Interest*, 109–27. Toomer stresses that the motives for Greaves's travel were to improve his linguistic skills and gain access to astronomical manuscripts in Arabic and Persian. G. J. Toomer, *Eastern Wisdome and Learning: The Study of Arabic in Seventeenth Century England* (Oxford: Clarendon Press, 1996), 127–42. In passing, Toomer mentions that Greaves and his mentor, Bainbridge, looked for these manuscripts and conducted observations in the works of Ptolemy (in Alexandria) and Hipparchus (in Rhodes) "to evaluate the ancient observations that had been transmitted" (138). Shalev suggests that Greaves appealed to ancient sources in order to standardize units of measure. Zur Shalev, "Measurer of All Things: John Greaves (1602–1652), the Great Pyramid, and Modern Metrology," *Journal of the History of Ideas* 63, no. 4 (2002): 555–75.

3. John Greaves, *Miscellaneous Works of Mr. John Greaves Professor of Astronomy in the University of Oxford*, ed. Thomas Birch (London: Printed by J. Hughes for J. Brindley, 1737). See also Thomas Smith, *Vita quorundam eruditissimorum et illustrium virorom: quorum nomina exftant in pagina fequenti* (London: apud D. Mortier, 1707); John Ward, *The Lives of the Professors of Gresham College* (London: Printed by J. Moore, 1740).

4. John Greaves, *Astronomica quaedam ex traditione Shah Cholgii Persae: una cum hypothesibus planetarum: item excreta quaedam ex Alfergani elementis astronomicis, et Ali Kustigii de terrae magnitudine et sphaerarum coelestium a terra distantiis: cum interpretatione Latina* (London, 1652), p. A. To this work, Greaves attached another geographic essay, *Binae tabulae geographicae une Nassir Eddini Persae, altera Vlug Beigi Tatari.*

5. See, for example, Marshall Clagett, "The Medieval Latin Translations from the Arabic of the Elements of Euclid, with Special Emphasis on the Versions of Adelard of Bath," *Isis* 44, no. 1–2 (1953): 16–42. Also on Peurbach and Cremona, see E. J. Aiton,

"Peurbach's Theoricae Novae Planetarum: A Translation with Commentary," *Osiris*, 2nd ser., 3 (1987): 4–43.

6. Greaves, *Astronomica quaedam*, preface, A.

7. Ibid.

8. Ibid. In Greaves's work discussed here, he attaches no importance to deciding which language is the primordial source for astronomical terms: Persian and Arabic used the same expressions. Ibid., A.3.

9. 'Ali Ibn Mahmud al-Kirmani, *Ma'asir-i Mahmud Shahi*, ed. Ansari Nurulhasan (Dihli: Danishgah-i Dihli, 1968).

10. George Saliba, "A Sixteenth-Century Arabic Critique of Ptolemaic Astronomy: The Work of Shams al-Din al-Khafri," *Journal for the History of Astronomy* 25 (1994): 15–38.

11. Greaves, *Binae tabulae*, B3.

12. Greaves's travels were meant to produce manuscripts, and he was able to purchase an "original manuscript of Abualfeda." Later in the preface to *Binae tabulae*, Greaves stresses that the "learned men" are mistaken in relying only on Abū al-Fidā and on just one translation of his works. He suggests that they consult other Arab writers and compare their work with the earliest text of Abū al-Fidā. Greaves, *Binae tabulae*, A3–B3.

13. Owen Gingerich, *An Annotated Census of Copernicus' De Revolutionibus (Nuremberg, 1543, and Basel, 1566)* (Leiden: Brill, 2002), 262.

14. Greaves, *Binae tabulae*, B2. Scholars showed that Thābit Ibn-Qurrā was a source for Copernicus's theories of precession and trepidation. See Kristian P. Moesgaard, "Thābit Ibn-Qurrā between Ptolemy and Copernicus," in *Avet, avec, après Copernic: La reprèsentation de l'univers et ses consèquences èpistèmologiques; XXXIe semaine de synthèse, 1–7 juin 1973* (Paris: Blanchard, 1975), 67–70.

15. Greaves, *Binae tabulae*, B2. Greaves excused Ptolemy's many mistakes in latitudes for locations far from Alexandria, but thought that Copernicus presented inexcusable absurdities. See also Greaves's letter to Ussher, "An Account of the Latitude of Constantinople and Rhodes," in Greaves, *Miscellaneous Works*, 2:370.

16. Greaves looked for Arabic sources of astronomy and relied on Arab observations taken at two famous observatories—for example, those given in the tables of Naṣīr al-Dīn al-Ṭūsī and Ulugh Beg. The aim was to determine precisely the latitudes and longitude of the major Near Eastern cities so as to make better maps. Ibid., 2:364–71.

17. Since the beginning of his career in astronomy—that is, since the early 1630s—Greaves had conducted observations of lunar eclipses in Oxford. In London, he recorded his observations of lunar eclipses in a manuscript entitled *Elementa omnium scientiarum, praesertim mathematicarum*. See Ward, *Lives of the Professors*, 149.

18. But Ussher disregards Arabic sources in his letter to Hartlib of November 12, 1639. Ri Parr, *A Collection of Three Hundred Letters: Written . . . James Usher, Late Lord Arch-bishop of Armagh . . . and others; as . . . Is. Vossius, Hu. Grotius a.o.* (London, 1686), no. 304, p. 623.

19. John Greaves, *Pyramidographia; Or, A Description of the Pyramids in Ægypt*, in his *Miscellaneous Works*, 1:9.

20. Ibid., 1:xiv–xvi.

21. Ibid., 1:10–11.

22. Greaves, *Pyramidographia*, 1:84.

23. Ibid., 1:86.

24. John Greaves, *A Discourse of the Roman Foot and Denarius*, in his *Miscellaneous Works*, 1:167–79.

25. William Laud, "Sermon before the King Charles's Second Parliament," in his *The Works of the Most Reverend Father in God, William Laud, D.D., Sometime Lord Archbishop of Canterbury* (Oxford, 1847), 1:75.

26. Greaves, *Discourse of the Roman Foot*, 1:169–70.

27. Ibid., 1:191.

28. Greaves, "Of the Denarius," in his *Miscellaneous Works*, 1:259.

29. Greaves, *Discourse of the Roman Foot*, 1:191.

30. Anthony Wood, *Athenæ Oxonienses . . .* (1689) (Oxford: For the Ecclesiastical History Society, 1848), vol. 2, col. 157.

31. Smith, *Vita quorundam eruditissimorum*, 7.

32. W. B. Patterson, *King James VI and I and the Reunion of Christendom* (Cambridge: Cambridge University Press, 1997), 41.

33. Patterson, *King James*, 46.

34. Ibid., 41.

35. Ibid., 115.

36. This statement about the Greek Orthodox Church was made in 1619 by De Dominis, Archbishop of Spalato, at that time close to James I. Most of the theologians supporting James I argued that the English Reformed Church was actually part of a tradition going back to the first four centuries of Christianity. Ibid., 213–19.

37. On Jesuits and Capuchin activities in the Muslim world, see Arnold Wilson, "History of the Mission of the Fathers of the Society of Jesus, Established in Persia by the Rev. Father Alexander of Rhodes," *Bulletin of the School of Oriental Studies* 3, no. 4 (1925): 675–706; Francis Richard, *Raphaël du Mans, missionnaire en Perse au XVIIe sicle* (Paris: Société d'histoire de l'Orient, L'Harmattan, 1995).

38. F. H. Marshall, "An Eastern Patriarch's Education in England," *Journal of Hellenic Studies* 40 (1926): 187–89.

39. John Christianson, *On Tycho's Island: Tycho Brahe and His Assistants, 1570–1601* (Cambridge: Cambridge University Press, 2000), 140–41.

40. Patterson, *King James*, 126; for Kepler's letter to James (1607), see also Adam Jared Apt, "The Reception of Kepler's Astronomy in England: 1590–1650" (Ph.D. diss., S.I., St. Catherine's College, 1982), 22. Kepler also wrote to James about harmony. See Johannes Kepler, *Gesammelte Werke*, ed. Max Caspar (Munich: Beck, 1954), xvi, 103–4.

41. Patterson, *King James*, 126. See also Apt, "Reception of Kepler's Astronomy," 23.

42. Laud, "Sermon before King Charles's Second Parliament."

43. William Laud, *True Relations of Sundry: Conference Had between William Laud, Late Lord Archbishop of Canterbury, and Mr. Fisher the Jesuit: by the Command of King James* (Oxford: John Henry Parker, 1849), 337.

44. Ibid., 427.

45. On Laud's collecting of Near Eastern manuscripts, see Feingold, "Patrons and Professors."

46. Patterson, *King James*, 217–19.

47. Toomer, *Eastern Wisdome and Learning*, 108–9.

48. In a letter from Constantinople, dated August 2, 1638, Greaves reports to Laud that he received "the expectations of *Clemens* [Clemens I, the first Bishop of Rome] with the *Catena of Job* [the text of Job surrounded on the margins with ancient commentaries], both [of] which, according to his Grace's instruction, I should have presented to the Patriarch." Greaves, *Miscellaneous Works*, 2:434. Indeed, ancient commentaries on the Book of Job might have been of interest to Laud, who perhaps, like della Valle, was interested in locating its ancient versions.

49. Greaves, *Miscellaneous Works*, 2:435.

50. John Jewel, *An Apology of the Church of England* (*Apologia*, 1562) (Ithaca, NY: Cornell University Press, 1963). See also Mason Southgate, *John Jewel and the Problem of Doctrinal Authority* (Cambridge, MA: Harvard University Press, 1962).

51. For praise of Jewel's work, see James Ussher, *Gravissimae quaestionis de Christianrum Ecclesiarum . . .* (Dublin: Hodges, Smith, 1864), 2:27.

52. James Ussher, *The Causes of the Continuance of the Contentions concerning the Church Government*, in *The Whole Works of the Most Rev. James Ussher*, ed. Charles Richard Elrington (Dublin: Hodges, Smith, 1864), 17:xix.

53. James Ussher, *Tractatus de controversiis pontificiis*, in Ussher, *Whole Works*, 14:41–46.

54. Ussher, *Whole Works*, 16:29, 248.

55. Letter CII, "Mr. Thomas Davis to the Archbishop of Armagh," in Ussher, *Whole Works*, 15:323–27. Ussher also distributed the copies he obtained to Laud and other colleagues. In the dedication to Laud on the Samaritan Pentateuch, he wrote: "a Samaritan Pentateuch, describes in ancient phonetics and letters the ecclesiastical writers." Ussher, *Whole Works*, 1:91.

56. Letter 127, in Ussher, *Whole Works*, 15:380–87.

57. Letter C, "Mr. Ralph Skinner to the Archbishop of Armagh," in Ussher, *Whole Works*, 15:320.

58. In 1644, Greaves wrote to Ussher: "The true solar year according to Chateans is 365 days and 2436/10000 parts of day. According to Nassir Eddin 365 days 14′32″30‴ According to Aly Kosgy, . . . 365 days 14′33″32‴, whereas Ptolemy is much more, 365 days 14′48″." Ussher, *Whole Works*, 7:73.

59. In *De Macedonum et Asianorum anno solari* (1648), Ussher endeavored "to restore the rules of finding the Macedonian year" to accommodate the ratio of the years in the Julian and Macedonian calendars. Ussher, *Whole Works*, 7:391.

60. James Ussher, *Geographical and Historical Disquisition touching the Asia Properly So Called*, in Ussher, *Whole Works*, 7:3.

61. John Greaves, "An Account of the Latitude of Constantinople, and Rhodes, Written by the Learned Mr. John Greaves, Sometime Professor of Astronomy in the University of Oxford, and Directed to the Most Reverend James Ussher Archbishop of Armagh," *Philosophical Transactions of the Royal Society of London* 15 (1685): 1295–1300 (reprint, New York: Johnson Reprint Corp.; Kraus Reprint Corp., 1963).

62. Ibid., 1298

63. Ibid., 1299,

64. See preface to John Greaves, *Epochæ celebriores, astronomicis, chronologicis, chataiorum, Syro-Græcorum, Arabum, Persarum, chorasmiorum, usitatæ, ex traditione Ulug Beigi, Indiæ cirta extraq; gangem principis, eas primus publicavit, recensuit, & commentariis illustravit, Johannes Gravius* (London, 1650).

65. Ussher, *Whole Works*, 7:74.

66. John Greaves, "Reflexions on a Report Made by Lord Treasurer Burleigh to the Lords of the Council . . . as Concerning the Needful Reformation of the Vulgar Kalendar, for the Civil Years and Days Accompting of Verifying According to the Time Truly Spent," *Philosophical Transactions of the Royal Society of London* 21 (1699): 343–54.

67. Edward Brerewood (1565–1613), first professor of astronomy at Gresham College, wrote on logic, linguistics, geography, optics, and weights and measures of the ancient world and was a member of the Society of Antiquaries. One of his works, *Enquiries touching the Diversities of Languages and Religions* (1614), published posthumously, dealt with corruption of the meanings of words and advocated a return to primordial sources of antiquity; this fit another agenda: Protestant unity. See Ward, *Lives of the Professors*, 74–75. See also Christopher Hill, *The Intellectual Origins of the English Revolution* (Oxford: Clarendon Press, 1965), 51.

68. John Greaves, *Iohannis Bainbrigii, astronomiae, in celeberrimâ Academiâ Oxoniensi, professoris Saviliani, Canicularia: unà cum demonstratione ortus Sirii heliaci, pro parallelo inferioris Aegypti* (Oxford, 1648).

69. Bainbridge's letter in Ussher, *Whole Works*, 15:213.

70. Apart from his linguistic interest in astronomy, as expressed in *Canicularia*, Bainbridge published other works of technical astronomy, such as *An Astronomical Description of the Comet of 1618* (London, 1619) and *Procli sphaera et Ptolmaei de hepothesibus planetarum* (London, 1620). In both, he adopted the new explanations of Tycho for the comet, and although he used Kepler as a source, he still did not admit Copernican cosmology. On Bainbridge and Kepler, see Apt, "Reception of Kepler's Astronomy," 188–200.

71. On Briggs's knowledge of mathematics, see Ward, *Lives of the Professors*, 120–28; Hill, *Intellectual Origins*, 38–58; Mordechai Feingold, *The Mathematicians' Apprenticeship: Science, Universities, and Society in England 1560–1640* (Cambridge: Cambridge University Press, 1984); Herman Goldstine, *A History of Numerical Analysis from the Sixteenth through the Nineteenth Century* (New York: Springer, 1977). On Briggs's cartography, see Katherine Neal, "Mathematics and Empire, Navigation and Exploration: Henry Briggs and the Northwest Passage Voyage of 1613," *Isis* 93, no. 3 (2002): 435–53.

72. Henry Briggs and Henry Gellibrand, *Trigonometria Britanica* (London: R. & W. Leybourn, 1658), 14. Briggs wrote the first book on logarithms, and Gellibrand completed and published the work after Briggs's death.

73. In the last ten years of their lives, Briggs and Kepler were in personal contact. In 1625, Briggs sent Kepler a copy of his *Arithmetica logarithmica* (1624). Briggs to Kepler, February 20–March 2, 1626, in Kepler, *Gesammelte Werke*, 9–11, 18, 220. On the first English Copernicans and the role Briggs played in the reception of Kepler in England, see Apt, "Reception of Kepler's Astronomy," 176–88.

74. At the convocation of parliament at Oxford in 1625, Thomas James, Bodleian Librarian, commissioned certain scholars to pursue the patristic manuscripts to detect the forgeries introduced by Roman Catholic editors. But, in fact, he had already started this work with the help of friends. In a letter to Ussher in February 1624, Thomas James described the progress of his work in comparing printed materials with ancient patristic manuscripts and stressed that he would need the help of Bainbridge and Briggs to obtain some manuscripts and compare them. In the letter, the librarian complains about the intrigues of the Roman Church and writes: "I have restored three hundred citations, and rescued them from corruption in thirty quire of paper: Mr. Briggs will satisfy you in this point, and sundry other projects of mine." Ussher, *Whole Works*, 15:266–67.

75. Ward, *Lives of the Professors*, 132.

76. Laud, *Works of the Most Reverend Father*, 5:84, 99, 163.

77. There was further reception of the works of Pocock and Greaves. Thanks to the sponsorship of Robert Boyle, in 1660 Pocock published an Arabic translation, with emendations and a new preface, of Grotius's tract *De Vertate religionis Christianae*, undertaken in the hope of converting Muslims. Pocock's *Bar Hebraeus, Specimen historiae Arabum* (1649) contains an extract of Abu al-Faraj's *Universal History* [*Mukhtasar fi al-duwal*], which functions as a template for a series of elaborate essays on Arabic history, natural philosophy, literature, and religion. This exploration became a chief source for later orientalists. Further, Pocock participated in the preparation of *Biblia sacra polyglotta*, and he published *Porta Mosis* (Oxford, 1655), which is the Judeo-Arabic edition of the prefatory discourses of Maimonides on the Mishnah, with Latin translation and notes. Pocock's *Specimen* and *Porta Mosis*, together with Greaves's work *Bainbrigii canicularia* (1648), were the first Arabic and Hebrew texts printed at the Oxford University Press.

78. This is not the famous William Gilbert (d. 1640), physician of Elizabeth I and writer on magnetism, but the William Gilbert who was probably born in 1597 and died December 18, 1640. He graduated from Cambridge University with a B.A. in 1616–17, M.A. in 1620, B.D. in 1627, and D.D. in 1632. See John Venn and J. H. Venn, eds., *Alumni cantabrigienses: A Biographical List of All Known Students, Graduates, and Holders of Office at the University of Cambridge, from the Earliest Times to 1900* (Cambridge: Cambridge University Press, 1622), pt. I, vol. 2, p. 215. Gilbert published *Architectonice consolationis: Or the Arat of Building Comfort . . .* (London, 1640).

79. Ussher, *Whole Works*, 16:41.

80. John Wilkins, *The Discovery of a World in the Moone; Or, A Discourse Tending*

to Prove, That 'Tis Probable There May Be Another Habitable World in That Planet (London: Printed by E.G. for Michael Sparke and Edward Forrest, 1638).

81. John Wilkins, *A Discourse concerning a New World & Another Planet* (London, 1640).

82. Wilkins, *Discovery of a World*, 18.

83. Ibid., 80.

84. See Shapin and Schaffer, *Leviathan and the Air Pump*, 106–7, 311–13.

85. See Barbara Shapiro, *John Wilkins, 1614–1672: An Intellectual Biography* (Berkeley and Los Angeles: University of California Press, 1969), 87. Although, in 1656, Wilkins married Robina, the sister of Cromwell, he was still attached to royalists such as John Evelyn and Hooke, who were friendly with Greaves. During the 1640s, Wilkins was involved in a gathering of philosophers in London that was called the "invisible college" of Robert Boyle. It eventually led to establishment of the Royal Society, in whose formation Wilkins was considered a central figure.

86. From the first page of the introduction, "To the Reader," in John Wilkins, *An Essay towards a Real Character and a Philosophical Language* (London: Printed for the Royal Society, 1668), emphasis added.

87. Wilkins, *Essay towards a Real Character*, 191.

88. Ibid., 191–92.

89. Amir Alexander, *Geometrical Landscapes: The Voyages of Discovery and the Transformation of Mathematical Practice* (Stanford: Stanford University Press, 2002), 73. On Briggs, see pp. 60–73.

90. See preface in Greaves, *Miscellaneous Works*, 1:A2.

91. See, for example, Toomer, *Eastern Wisdome and Learning*, 127–34.

92. Feingold, *Patrons and Professors*, 109–27.

93. Isaac Newton, *A Dissertation upon the Sacred Cubit of the Jews and the Cubits of the Several Nations in Which, from the Dimensions of the Greatest Egyptian Pyramid, as Taken by Mr. John Greaves, the Antient Cubit of Memphis Is Determined*, in Greaves, *Miscellaneous Works*, 2:405.

94. Ibid., 408, 417–18, 419.

95. Ibid., 425, 426.

96. In the introduction to *Opticks*, Newton states that Noah was the last to hold the perfect knowledge about nature. Isaac Newton, *Opticks; Or, A Treatise of the Reflexions, Refractions, Inflexions and Colours of Light* (London: Printed for S. Smith and B. Walford, 1704).

Five · Exchanging Heavens and Hearts

1. Noël Duret, *Nouvelle théorie des planètes: Conforme aux obseruations de Ptolomée, Copernie, Tycho, Lansberge, & autres excellens Astronomes, tant anciens que modernes. Avec les tables Richeliennes et Parisiennes, exactement calculées* (Paris, 1635).

2. Ibrāhīm Efendi al-Zigetvari Tezkireci, *Sajanjal al-aflāk fi ghāyat al-idrāk*, Kandilli Rasthanesi Collection, MS 403, Istanbul. For more than three centuries, this manuscript lay buried in an Ottoman archive in Constantinople—until it resurfaced

in the early 1990s. In reacting to the well-entrenched assumption that Islamic natural philosophy stagnated, especially during the Ottoman period, the modern Turkish scholar Ekmeleddin İhsanoğlu has gone a long way toward providing an argument that Islamic natural philosophy had its own Ottoman flowering, one that arose within a creative awareness of the "Scientific Revolution." His cataloguing of scientific manuscripts shows that "Ottoman science" closely followed European innovations and also progressed as a separate offshoot of European developments. İhsanoğlu argues that, from early on, Ottoman astronomers had been interested in the Copernican system and that, despite the rise of mysticism, certain positivist islands of astronomy survived. Ekmeleddin İhsanoğlu, "Introduction of Western Science to the Ottoman World: A Case Study of Modern Astronomy (1660–1860)," in *The Transfer of Modern Science and Technology to the Muslim World*, ed. Ekmeleddin İhsanoğlu (Istanbul: Research Centre for Islamic History, Art, and Culture, 1992). See also Ekmeleddin İhsanoğlu, ed., *Osmanlı astronomi literatürü tarihi* [History of Astronomical Literature during the Ottoman Period] (Istanbul: Research Centre for Islamic History, Art, and Culture, 1997), 1:340–45; Ekmeleddin İhsanoğlu, "The Introduction of Western Science to the Ottoman World: A Case Study of Modern Astronomy (1660–1860)," in *Religious Values and the Rise of Science in Europe*, ed. John Brooke and Ekmeleddin İhsanoğlu (Istanbul: Research Centre for Islamic History, Art, and Culture, 2005), 185–228.

3. Noël Duret, *Novæ motuum cælestium ephemerides Richelianæ* . . . (Paris, 1638).

4. In *Biographie universelle: Ancienne et moderne* (Paris: Michaud frères, 1811–62), 12:100, we read that Noël Duret "taught mathematics in Paris, and published a few books, none of which met with any real success." Jèrôme de la Lande mentioned Duret's works, and referred to Duret as a "professor of sciences and mathematics," in his *Bibiographie astronomique: avec l'histoire de l'astronomie depuis 1781 jusquà 1802* (Paris, 1803), 205, 206, 208, 210, 212, 224. La Lande was one of those first post-Copernican European astronomers whose texts, like Duret's, were translated into Arabic. See Ḥusayn Ḥusnī, al-munajjim al-thānī, *Tahdhīb al-anām fī ta'arīb zīj Lalande nam* [an Arabic translation, rectified, of la Lande's astronomical map] (ca. 1850), catalogue no. 2.2.17, Dār al-Kutub al-Miṣriyya, Cairo.

5. Philips Lansbergen, *Tabulae motuum coelestium perpetuae; ex omnium temporum observationibus constructae, temporumque omnium observationibus consentientes: Item Novae et genuinae motuum coelestium theoricae & Astronomicarum observationum thesaurus* (Middelburg, Holland, 1632).

6. Noël Duret, *Supplementi tabularum Richelienarum, pars prima, cum brevi planetarum, Theoria ex Kepleri, sententia* (Paris, 1639). The version published in London in 1647 had a binding with the title *Astrology*. Books at this time were sold as stacks of paper without a publisher's binding, so the *Astrology* title was simply idiosyncratic to that particular bookseller. The work was published by John Benson as Noël Duret, *Novæ motuum cælestium ephemerides Richelianæ annorum 15, ab anno 1637 incipientes, ubi sex anni priores e fonti bus Lansbergianis bāshī, reliqui vero e numeris Tychoni-Keplerianis eruntur, quibus accesserunt. In priori parte. 1 Isagoge in*

Astrologiam. 2 De aeris mutatione. 3 Doctrina primi mobilis exquisite demonstrata. In secunda parte. 1 Usus tabularum astronomicarum pro rebus omnibus ad astronomiam spectantibus instituendis. 2 De crisium mysterio tractatus. 3 Gnomonices liber unus, ubi scioterica delineandi horologia quocunque modo vel declinantia, vel inclinantia methodus omnium & facillima & brevissima tabularum ope, traditur. Authore Natal. Durret, cosmographo Regio, at eminentiss (London, 1647).

7. A dictionary of Ottoman biography, *Sicill-i Osmanî* (Beşiktaş, Istanbul: Kültür Bakanlığı ile Türkiye Ekonomik ve Toplumsal Tarih Vakfı'nın ortak yayınıdır, 1996) has many entries for "Ibrāhīm Efendi," but none referring to a "Tezkireci" or to a person interested in astronomy and astrology. Biographical dictionaries of Arabic writers are similarly unhelpful.

8. "Mr. Flamsteads letter of July 24 1675 to the Publisher, relating to Another, Printed in Num. 110 of These Tracts, concerning M. Horroxes Lunar System," *Philosophical Transactions of the Royal Society of London* 10 (1675): 370. See also the letter from the mathematician Abraham Sharp to John Flamsteed, dated March 23, 1704/05, in *The Correspondence of John Flamsteed, the First Astronomer Royal* (Bristol and Philadelphia: Institute of Physics Publishing, 2002), 3:149, letter no. 991. Curtis Wilson mentions Flamsteed as an anti-Copernican astronomer who was interested in Kepler's elliptical model as an explanation for the motion of the heavenly bodies, without accepting the grand scheme of heliocentrism. Curtis Wilson, "Kepler's Derivation and Elliptical Path," *Isis* 59, no. 1 (1968): 22, n. 88. Riccioli, in *Amagestrum novum*, notes that Noël Duret, a maker of ephemerides who had previously followed Lansbergen's tables, was so impressed by Gassendi's confirmation of the Keplerian prediction that he deserted to the Rudolphine camp. See Curtis Wilson, "From Kepler's Laws, So-Called to Universal Gravitation: Empirical Factors," in his *Astronomy from Kepler to Newton* (London: Variorum, 1989), 100, n. 29.

9. Nicholas Culpeper, *Semiotica urinica: Astrological Judgment of Diseases*, 4th ed. (London: Nathaniell Brookes, 1671), 13. The connection between medicine and astrology in Islam was made early on, at the beginning of the move toward translating Greek philosophy. See Yuhanna ibn Bahtisu', *Kitāb fima Yahtagu ilaihi al-tabīb min 'ilm al-nujūm* [What the Physician Ought to Know about Astrology] (ca. 10th century), and Yuhanna Ibn al-Salt, *Kitab ṭibb nujūmī* [Book of Astrological Medicine], both in *Astromathematics in Islam*, ed. Felix Klein-Franke (New York: Georg Olms Verlag, 1984). An important practitioner of astrological medicine was the German-Swiss physician Paracelsus Theophrastus von Hohenheim (1493–1541). His ideas were based on analogies between macrocosm and microcosm: human organs were thought to correspond to celestial bodies—for example, the heart to the sun. Furthermore, Hohenheim and his followers preceded Copernicus in emphasizing the centrality of the sun. This may explain the emphasis in al-Zigetvari's introduction to his translation of Duret on the translator's interest in the heliocentric system.

10. Culpeper, *Semiotica urinica*, 1. Culpeper's "Avenezra" was also known as Avraham Ibn-'Ezra, the twelfth-century Jewish hermeneutic scholar of the Bible, as well as poet, grammarian, traveler, neo-Platonic philosopher, astrologer, and astronomer,

whom we have encountered in earlier chapters. Ibn-'Ezra was best known as a biblical exegete whose commentaries (e.g., Ex. 23:28, 33:21; Eccl. 3:1 and 7:13) contributed to the golden age of Spanish Judaism. In his astrological writings, he claimed that the Jewish Temple and its instruments resembled the cosmos. On Ibn-'Ezra's debate with the Muslim astrologer Abū-Ma'shar (Albumasar), see Ibn-'Ezra's *Otsar haHaim* [The Treasure of Life], in Culpepper, *Semiotica urinica*, 45. On Ibn-'Ezra's advocacy of Ptolemaic astronomy and opposition to Ptolemaic astrology, see his *Sefer haTe'amim* [The Book of Reasons], in Culpeper, *Semiotica urinica*, 51.

11. Culpeper, *Semiotica urinica*, title page.

12. Duret, *Novæ motuum cælestium ephemerides Richelianæ* (1638), 13. Duret was not the only protégé of Richelieu engaged in astronomy and astrology. His colleague Spinelli Davide wrote a historical satirical work dedicated to Cardinal Richelieu; it is divided into five dialogues, the last of which, devoted to astrology and astronomy, includes an astronomical diagram. Spinelli Davide, *Giove appresso gli Etiopi* (Venice, 1633).

13. Duret, *Novæ motuum cælestium ephemerides Richelianæ* (1638), dedication to Richelieu. Duret also writes in the dedication that the tables "ought not be inscribed with my name, since I would not have been able either to take up this Herculean task (never before attempted by any Frenchman) or to carry it through to the end if I were lacking Your Eminence's customary generosity, under which I so advanced with continuous labor and sleepless nights, setting aside my household duties and neglecting important business matters." From this we can deduce that the astronomical tables—which were not a completely original work and relied on a variety of sources, especially the Rudolphine tables—were included merely to anchor the work in Paris. Richelieu's patronage arose from a motive to match the achievements of the Habsburg court and to make Paris a new focus for astrological practitioners from varying locales. More importantly, it made the meridian of Paris the reference point for expeditions that sailed to measure longitudes, especially in the New World. This represents a case of astronomy's being called upon to solidify an empire's political self-awareness and to give Richelieu, in Paris, a new display of power. Duret compiled the new Paris-based tables with reference to other fields of science—that is, not solely astronomy, but mathematics, geography, hydrography, and geodesy—so that the tables would have practical use in bureaucracies working for the "public benefit." He also hoped to facilitate the charting of the earth in reference to the position of the stars, "not just by measuring time and hours and the parts of hours without error, but also, by a similar system, to perceive on the Earth the signs of the stars, in such a way that with little difficulty we may discern everything that is gathered in the vast and immense sky, not only of the sun, but also of the rest of the stars that are contained in the circuit of heaven—all of this as we stand on the Earth as if on a point." However, the book's main purpose concerns medical astrology: "It is the mark of a prudent doctor to foresee threatening storms of diseases from the contemplation of the heavens, as if from a looking-glass. He sails without rudder and oar, eventually to wreck his ship, who practices medicine without any observation of the seasons and

the stars, for astrology is the doctor's eye, and he who is lacking this will deservedly be called 'blind.'" At the end of the dedication, Duret mentions that he owed a personal debt to Richelieu and declares that "I reverently offer to Your Eminence this useful little work, arising from the Parisian Richelieuan Tables," like a promised debt, and ". . . I beg as a suppliant that you deem this work of many late nights worthy of being received with a kind expression, O greatest of Dukes . . . I faithfully pray to God, the Best and the Greatest, that he prolong Your Eminence's life and good health for many years to come, and that he completely bestow the profit of your tables." Duret, *Novæ motuum cælestium ephemerides Richelianæ* (1638), dedication.

14. In the late sixteenth century, France and the Ottomans formed an alliance to counter the Habsburgs and the Spanish, and the Ottomans established a naval station at the French port of Toulon. See Marshall Hodgson, *The Venture of Islam: The Gunpowder Empires and Modern Times* (Chicago: University of Chicago Press, 1974), 117. It seems that this alliance was not strictly military, and G. R. R. Treasure tells us about Richelieu's "dream" of extending Mediterranean trade to the East, as a counterbalance to the rise of the Atlantic trade dominated by England, Spain, and Holland—all enemies of France. G. R. R. Treasure, *Cardinal Richelieu and the Development of Absolutism* (London: A. and C. Black, 1972), 210–11. Robert Knecht mentions Richelieu's great interest in opening up eastern trade routes for French merchants, for which purpose he established personal relations and mediated through envoys (mainly Harlay de Cèsy) with the Ottoman court. Robert Knecht, *Richelieu* (London: Longman, 1991), 160.

15. Natalie Zemon Davis mentions the exchange of ambassadorial gifts between the French king and the Ottoman sultan and suggests that "Ottoman objects and manuscripts were sought in the west as signs of beauty, curiosity, and wished-for domination." Natalie Zemon Davis, *The Gift in Sixteenth-Century France* (Madison: University of Wisconsin Press, 2000), 126. For lists of gifts exchanged between the French and the Ottoman courts, see Clarence Dana Rouillard, *The Turk in French History, Thought, and Literature (1520–1660)* (Paris: Boivin, 1940), 87; 134. n. 2; 165. Carl Burckhardt provides some examples. Gentile Bellini, whom Signoria of Venice had sent to Istanbul in 1479, made the famous portrait of the Sultan Mehmed II. The son of Mehmed II, Sultan Bayezid II (1481–1512), received a letter containing plans for a windmill, a ship's pump, and a large drawbridge to span the Bosporus (the letter was written by Leonardo da Vinci, whom Bayezid II had tried to engage as a bridge-builder in 1506). In 1525, Francis I of France sent a letter to Süleyman the Magnificent in which he asked for help against the Habsburgs. The letter resulted in the alliance between France and the Ottomans, which is described by Burckhardt as "the blasphemous union of the lily and the crescent." It led to the conquest of parts of Hungary by the Ottomans, including—most importantly for identifying al-Zigetvari—the city of Szigetvar. It was while he was leading the siege of Szigetvar that Süleyman died. In Richelieu's time, Louis XIII was the only European monarch referred to by the Ottoman sultans as "Padishah," the emperor. Carl Burckhardt, *Richelieu and His Age* (London: Allen & Unwin, 1970), 2:284–95.

16. On gift exchange, see Pierre Bourdieu, *Outline of Theory of Practice*, trans. Richard Nice (Cambridge: Cambridge University Press, 1977), 5–6, 38,109; Natalie Zemon Davis, "Beyond the Market: Books as Gifts in Sixteenth-Century France," *Transactions of the Royal Historical Society*, ser. 5, 33 (1983): 69–88; Sharon Kettering, "Gift-Giving and Patronage in Early Modern France," *French History* 2 (1988): 133–51.

17. See *Richelieu et le monde de l'esprit Sorbonne: 1985* (Paris: Impr. Nationale, 1985), items 40, 26, 249, 250, 251.

18. The most significant of these manuscripts is *Kitāb ta'alīm al-Masīḥī* [Book of Instructions to the Christian], which is a translation of Richelieu's *Instruction du chrestien*. The translation was made at the beginning of the seventeenth century by Jester de Paussi, the head of the Capuchins of Baghdad.

19. De Breves was much appreciated by Richelieu for convincing the Ottomans to issue the proclamation of capitulations that enabled French merchants and Capuchin missionaries to move freely in the Ottoman Empire. François Savary de Breves, *Seigneur François Savary, Articles du traité fait entre Henri le Grand, roi de France & de Navarre; & Sultan Amat, empereur des Turcs, en l'année 1604: Par l'entremise de messire François Savary, seigneur de Breves, conseiller du roi en ses conseils d'etat & privé, lors ambassadeur pour sa majesté à la porte dudit empereur. L'empereur Amat fils de l'impereur Mehemet toûjours victorieux* (Paris, 1604). In Paris, de Breves worked on developing a press to print Arabic works to sell in the Ottoman Empire. See the entry on de Breves in *Biographie universelle*.

20. In his fondness for art, Richelieu never hesitated to call on subordinates to exploit the apparatus of the absolutist state for his personal benefit. De Breves used his foothold in Constantinople to procure Islamic artistic and scholarly objects and to deliver gifts to the Ottomans in exchange, gifts of artistic, scientific, and technological interest. After de Breves's death in 1628, de Cèsy managed the process of gift exchange. For sources of France-Ottoman gift exchange and the role of de Cèsy, see Gèrard Tongas, *Les relations de la France avec l'empire ottoman durant la première moitiè du xviie siècle et l'ambassade à Constantinople Philippe de Harlay, comte de Cèsy, 1619–1640* (Toulouse: Impr. F. Boisseau, 1942).

21. For de Cèsy's papers, see ibid., 35. I suspect that the occasion of de Cèsy's presenting Duret's book was a meeting with the Sultan in 1638, in which de Cèsy negotiated in the name of Richelieu for approving passage of Kazak troops through the Ottoman-Hungarian province on their way to attack the Habsburgs as a rear action.

22. Nathan Sivin describes how, in China, the Copernican system was presented through the Jesuits' particular reading of Tycho. Nathan Sivin, "Copernicus in China," *Studia Copernicana* 6 (1973): 63–122. Unlike the Jesuits in China, Duret emphasized Tycho for his astronomical model and Kepler for his astrology, mathematics, and mysticism: Copernicanism was marginal. For an account of the new astronomy in China, see Keizo Hashimoto, *Hsü Kuang-Ch'I and Astronomical Reform: The Process of the Chinese Acceptance of Western Astronomy, 1629–1635* (Osaka: Kansai University Press, 1988).

23. Annemarie Schimmel, *Mystical Dimension of Islam* (Chapel Hill: University of North Carolina Press, 1975), 379.

24. Gustav Bayerle, *Pashas, Begs, and Effendis: A Historical Dictionary of Titles and Terms in the Ottoman Empire* (Istanbul: Isis Press, 1977), 38, 126–27, 147.

25. Burckhardt, *Richelieu and His Age*, 288.

26. Aulijā Çelebī, Robert Dankoff, and Klaus Kreiser, eds., *Evliya Çelebi's Book of Travels: Land and People of the Ottoman Empire in the Seventeenth Century; a Corpus of Partial Editions* (Leiden: Brill, 1990), 3:56–60.

27. For examples of Hungarians having a career in the Ottoman army and bureaucracy, see Fodor Pál, "Making a Living on the Frontiers: Volunteers in the Sixteenth-Century Ottoman Army," and Gèza Dávid, "An Ottoman Military Career on the Hungarian Borders: Kasim *Voyvoda*, Bey and Pasha," both in *Ottomans, Hungarians, and Habsburgs in Central Europe: The Military Confines in the Era of Ottoman Conquest*, ed. Gèza Dávid and Fodor Pál (Leiden: Brill, 2000), 229–64, 265–98.

28. Al-Zigetvari, *Sajanjal*. The word for "purpose," rendered as *ghāyāt* in Arabic and *picatrix* in Latin, had a special resonance in both cultures. An Arabic book entitled *Ghāyat al-Ḥakīm* (The Purpose of the Wise Man) was supposedly the work of the Muslim Spanish mathematician Abu al-Qāsim Maslama al-Maghītī, translated into Latin during the thirteenth century at the request of Alfonso I of Castile, with the title *Picatrix*. Thereafter, writers in Arabic and Latin frequently mentioned the book, sometimes as a work dealing with blasphemous magic and sometimes as containing esoteric wisdom. Its subjects included cosmology and astrology. Possibly al-Zigetvari's selection of the word *ghāyāt* suggested a debt to *Ghāyat al-Ḥakīm*, which he could have read in its more popular, Latin version. Thus, his interest in Duret's book would have been driven by a previous fascination with magic, astrology, and cosmology. See the bilingual Arabic-German version, *Ghāyat al-ḥakīm wa-aḥaqq al-natījatayn bi-aītaqdīm = Picatrix, Das Ziel des Weisen* (Berlin: B. G. Teubner, 1933). See also David Pingree, ed., *Picatrix: The Latin Version of the Ghāyat al-Hakim* (London: Warburg Institute, 1986). For a more recent study, see Elizabeth Carnell, "Talismans and the Stars: Varieties of Interpersonal Magic in Ghayat al-hakim" (Ph.D. diss., University of Michigan, 2002).

29. See, for example, Orhan Pamuk, *The White Castle* (New York: Braziller, 1991), 59–60. In this novel, based on an autobiographical manuscript of a Christian captive in Istanbul, a servant of an Ottoman astronomer in the late seventeenth century, we find references to mirrors as not only reflecting physical bodies but also playing an epistemological role of self-reflection. By gazing at the mirror, Ottoman scholars were attempting to know themselves.

30. Mevlānā Jelāleddīn Rūmī, *Mesnevī*, in *Ottoman Lyric Poetry: An Anthology*, ed. Walter Andrews (Austin: University of Texas Press, 1997), 121.

31. Walter Andrews claims that "in these very particular ways we cannot really know Ottoman poetry without knowing something of Rūmī." Ibid.

32. H. Corbin, *Surawardi d'Alep, fondateur de la doctrine illuminative* (Paris: G. P. Maisonneuve, 1939).

33. Shihāb al-dīn Suhrawardī, *Ḥikmat al-ishrāq* [The Philosophy of Illumination], trans. John Walbridge and Hossein Ziai (Salt Lake City, UT: Brigham Young University Press, 1999), xi–xliii; *Encyclopedia of Islam* (Leiden: Brill, 1993), 783.

34. John Walbridge, *The Science of Mystic Lights: Qutb al-Din Shirazi and the Illuminationist Tradition in Islamic Philosophy* (Cambridge, MA: Harvard University Press, 1992), 35.

35. Suhrawardī, *Ḥikmat al-ishrāq*, 4.

36. For the philosophy of illumination of Avicenna, see Dimitri Gutas, *Avicenna and the Aristotelian Tradition* (Leiden: Brill, 1988), 115–30; Dimitri Gutas, "Avicenna's Eastern ('Oriental') Philosophy: Nature, Contents, Transmission," *Arabic Sciences and Philosophy: An Historical Journal* 10 (2000): 159–80.

37. Sohravardi [Suhrawardī], *The Book of Radiance: Partaw'namah: A Parallel English-Persian Text*, trans. and ed. Hossein Ziai (Costa Mesa, CA: Mazda, 1998), 81.

38. A. I. Sabra, *Optics, Astronomy, and Logic: Studies in Arabic Science and Philosophy* (Brookfield, VT: Variorum, 1994), 130–40.

39. al-Ḥassan Ibn Ḥusayn Ibn al-Haytham, *al-Shukūk 'alā Baṭlamyūs (Dubitationes in Ptolemaeum; Aporias against Ptolemy)*, ed. A. I. Sabra and N. Shehaby (Cairo: Dār al-Kuttub, 1971), 39.

40. Walbridge, *Science of Mystic Lights.*

41. There are different versions of these Sufi practices. Briefly, Turkish and Persian mystical poetry of the late medieval period emphasized the practice of *dhikr qalbī* (recollection for distilling the heart) by which the pure mirror reflects the heavens and God and symbolizes the merging of the believer with God and nature. See Schimmel, *Mystical Dimension*, 169–75, 188–90, 228.

42. A. J. Arberry, *Discourses of Rumi* (London: J. Murray, 1961), 62–63.

43. Taqī al-Dīn Muḥammad Ibn-Ma'ārūf, *Kitāb al-kawākib al-durrīyah fī bānkāmāt al-dawrīyah* [The Revolving Planets and the Revolving Clocks], Mīqāt Collection, MS 557/1, p. L.4a, Dār al-Kuttub, Cairo. For a printed commentary (although not precise in translation), see Sevim Tekeli, *16'ıncı asırda Osmanlılarda saat ve Takiyüddin'in Mekanik saat konstrüksüyonuna dair en parlak yıldızlar adlı eseri: The Clocks in Ottoman Empire in 16th Century and Taqī al-Dīn's "The Brightest Stars for the Construction of the Mechanical Clock"* (Ankara: Ankara Universitesi Basimevi, 1966)

44. Taqī al-Dīn, *al-Kawākib al-durrīyah fī bānkāmāt al-dawrīyah*, L.4a.

45. Annemarie Schimmel, "Calligraphy and Sufism in Ottoman Turkey," in *The Dervish Lodge: Architecture, Art, and Sufism in Ottoman Turkey*, ed. Raymond Lifchez (Berkeley and Los Angeles: University of California Press, 1992), 242–53.

46. The 1650s and 1660s were characterized by political and cultural turmoil in Istanbul that originated in the struggles among different Sufi orders and to some extent between these orders and the orthodox clergy. See Derin Terzioğlu, "Sufi and Dissident in the Ottoman Empire: Niyāzī-i miṣrī" (Ph.D. diss., Harvard University, 1999), 190–277.

47. Muḥammad Ibn al-Zubayr and al-Sa'īd Muḥammad Badawī, *Sijill asmā' al-'Arab* (Beirut: Maktabat Lubnān, 1991), 2:287.

48. Collins presents an illuminating diagram representing the connections between mystics, scientists, and logicians in late medieval Islamic natural philosophy, in which most of the central philosophers were somehow in touch with illuminist philosophy or mysticism. Randall Collins, *The Sociology of Philosophies: A Global Theory of Intellectual Change* (Cambridge, MA: Harvard University Press, 1998), 424, fig. 8.3. See also Naṣīr al-Dīn al-Ṭūsī, *Zubdat al-idrāk fī* hay'at al-aflāk: ma'a dirāsat al-manhaj al-Ṭūsī al-'ilmī fī majāl al-falak, ed. 'Abbās Muĥammad Muḥammad Ḥasan Sulaymān (Alexandria: Dār al-Ma'rīfah al-Jāmi'īyah, 1994).

49. I am trying to show here that the most prominent astronomer, al-Ṭūsī, who has been discussed by historians of Islamic science mainly for his astronomy and mathematics, was an intellectual and philosopher of much wider scope. Astronomy was, for al-Ṭūsī, part of a larger project to acquire unified knowledge. Hence, I argue that he was aware of the shortcomings inherent in perception through reason or senses and looked for other kinds of perception (intuitive). In some of his works, al-Ṭūsī Persianized Arabic prose by assimilating it into an eloquent and fluent Persian diction, following the model of Suhrawardī. See Hamid Dabashi, "Khawājah Naṣīr al-Dīn al-Ṭūsī: The Philosopher/Vizier and the Intellectual Climate of His Time," in *History of Islamic Philosophy*, ed. Seyyed Hossein Nasr and Oliver Leaman (New York: Routledge, 1996), 1:239; Muhammad Mu'in, "Naṣīr al-Dīn al-Ṭūsī wa zabān wa adab pārsī," *Majalla yi danishkada-yi adabiyyāt* 3, no. 4 (1956): 30–42. In his commentary on Avicenna's *al-Ishārāt wa al-tanbīhāt*, al-Ṭūsī occasionally takes issue with Avicenna and prefers the positions of Suhrawardī.

The questions of ultimate knowledge and how to acquire it were intertwined in al-Ṭūsī's writings. In his autobiography, *al-Siyar wa-al-sulūk* (Contemplation and Action), he approached the problem through an Avicennian lens. Perfect knowledge is the knowing of a thing in which essence and existence, or potentiality and actuality, overlap—that is, God. In nature, the two are divided, so apperception of natural motion from potential to actual would end in absolute knowledge. The way to acquire it is not by spiritual or physical knowledge alone but by combining the two. See Naṣīr al-Dīn al-Ṭūsī, *Contemplation and Action: The Spiritual Autobiography of a Muslim Scholar [Sayr wa sulūk]*, ed. and trans. S. J. Badakhchani (London: I. B. Tauris, in association with Institute of Ismaili Studies, 1998), 25, 26, 29, 42, 43. Al-Ṭūsī had other channels that exposed him to the philosophy of illumination. In correspondence with the mystical philosopher al-Qūnawī, who was influenced by Suhrawardī, al-Ṭūsī expressed his understanding of the concept of *idrāk* and its connection to astronomy. Once again, he looks at movement from potentiality to actuality as the ultimate understanding of nature. Because this change occurs within time, and time is measured and determined by the motion of celestial bodies, the attainment of ultimate knowledge occurs through apperception of the souls of the stars and the interaction of the stars with the active intellect, both of which move the celestial bodies. Hence, *idrāk* aims at attaining knowledge of the sources of celestial bodies as well as the ways in which they move—a perfect knowledge. See Naṣīr al-Dīn al-Ṭūsī, *al-Murāsalāt bayna Ṣadr al-Dīn al-Qaūnawī wa-Naṣīr al-Dīn al-Ṭūsī* (Beirut: Yuṭlabu min Dār al-Nashr

Frānts Shtāynar, Shtūtgārt, 1995), 107–8, 112, 116–17. Therefore, unlike F. J. Ragep, I would argue that al-Ṭūsī's other fields of knowledge had a profound connection with his rational, mathematical astronomy. F. J. Ragep, *Naṣīr al-Dīn al-Ṭūsī's Memoir on Astronomy (al-Tadhkirat fi 'ilm al-hay'a)* (New York: Springer-Verlag, 1993), 1:9–23.

50. For an anecdote on a meeting between the famous Sufi poet Rūmī and Shīrāzī, the illuminist philosopher and astronomer, see Walbridge, *Science of Mystic Lights*, 14–15.

51. Ghiyāth al-Dīn Jamshīd al-Kāshī, *Uluğ Bey Ve Semarkanddeki Ilim Faaliyeti Hakkinda Giyasüddin Kāşi'nin Mektubu = Ghiyâth al Dîn al Kâshî's letter on Ulugh Bey and the Scientific Activity in Samarqand*, ed. Aydin Sayili (Ankara: Türk Tarih Kurumu Basımevi, 1960), 107.

52. Kātib Çelebī, *Sullam al-ūsūl ilā tabakāt al-fuhūl*, Shehid Ali Pasha Section, MS 1887, p. 271a, Süleymaniye Library, Istanbul.

53. Kâtip [Kātib] Çelebi, *The Balance of Truth* [*Mīzān al-ḥaqq*], trans. G. L. Lewis (London: Allen and Unwin, 1957), 25.

54. Ibid., 135–55.

55. In the Ottoman court, the *munajjim bāshī* was responsible for drawing up the horoscope for court activities and policies. Grammatically, the term is the active participle of *najjama*, "to observe the stars and deduce from them the state of the world." For a long time, the term as a noun designated both astrologer and astronomer, so close were the functions of the two. Astrology was also *'ilm al-ahkām*.

56. Al-Zigetvari, *Sajanjal*, 1–3. One can rely on Ekmeleddin İhsanoğlu's translation of the introduction, with only a few qualifications; in the text I give a slightly modified translation. Ekmeleddin İhsanoğlu, "Introduction of Western Science" (1992), 71–72, 84.

57. A. I. Sabra, "Configuring the Universe: Aporetic, Problem Solving, and Kinematic Modeling as Themes of Arabic Astronomy," *Perspectives on Science* 6, no. 3 (1998): 13.

58. Al-Zigetvari, *Sajanjal*, 2. In this extract, he mentions a work by Copernicus from 1525, which is an incorrect date. We know that sometime between 1507 and 1515, Copernicus completed a draft of a short astronomical essay, *De hypothesibus motuum coelestium a se constitutis commentariolus* (known as the *Commentariolus*), which was not published until the nineteenth century. Later, Copernicus worked on *De revolutionibus orbium coelestium* [On the Revolutions of the Celestial Spheres]. This was first published by a Lutheran printer in Nuremberg, just before Copernicus's death in 1543.

59. Thomas Kuhn, *The Copernican Revolution* (Cambridge, MA: Harvard University Press, 1956), 184.

60. Noel Swerdlow, "The Derivation and First Draft of Copernicus's Planetary Theory: A Translation of the *Commentariolus* with Commentary," *Proceedings of the American Philosophical Society* 117 (1973): 423–512. For the historiographic debate, see Robert Westman, "Two Cultures or One? A Second Look at Kuhn's *The Copernican Revolution*," *Isis* 85, no. 1 (1994): 90–93.

61. Some scholars drew connections between Copernicus and Thābit Ibn-Qurra (d. 901), who posed the notion of trepidation. See Kristian P. Moesgaard, "Thābit Ibn-Qurra between Ptolemy and Copernicus," in *Avet, avec, après Copernic: La représentation de l'univers et ses conséquences épistèmologiques; XXXIe semaine de synthèse, 1–7 juin 1973* (Paris: Blanchard, 1975), 67–70.

62. Robert Westman, "The Melanchthon Circle, Rheticus, and the Wittenberg Interpretation of the Copernican Theory," *Isis* 66 (1975): 167.

63. Al-Zigetvari, *Sajanjal*, 72.

64. *Tarjama Muqaddimāt Wâhiya fi Ahwāl al-Nujūm zū Zuāba*, Hafid Efendi Collection, MS 180, Süleymaniye Manuscript Library, Istanbul. The translator is unknown, but may have been Hoja Ishaq Effendi (d. 1834); the date of copying is nineteenth century. More information on this is available in İhsanoğlu, *Osmanlı astronomi literatürü tarihi*, 2:757–58.

65. This neglect is now being corrected. See an anthology of conference papers: F. J. Ragep and S. Ragep, eds., *Tradition, Transmission, Transformation: Proceedings of Two Conferences on Pre-Modern Science Held at the University of Oklahoma* (Leiden: Brill, 1996).

66. Aydin Sayili blamed the decline of Ottoman science on a failure to reconcile science with both philosophy and religion. Instead, "the Muslims hesitated to assume that natural processes were conducted according to certain principles." For Muslims, "there was a continuous interference of God, and of supernatural forces in nature . . . hence a kind of intellectual stupor." Aydin Sayili, "The Causes of the Decline of Scientific Work in Islam," in *The Observatory in Islam and Its Place in the General History of the Observatory*, ed. Aydin Sayili (Ankara: Türk Tarih Kurumu Basımevi, 1960), 422.

67. Following this erroneous approach, Seyyed Hossein Nasr, for instance, divides the scientific schools of Islamic philosophy into two dominant trends: first, the hermetic-Pythagorean school, which relied on symbolic interpretations of nature, commonly through mathematics, and second, an emphasis on rationalism, encyclopedism, and such figures as Aristotle, Ptolemy, and Galen. Seyyed Hossein Nasr, *Science and Civilization in Islam* (Cambridge, MA: Harvard University Press, 1987), 33.

Conclusion · From "Incommensurability of Cultures" to Mutually Embraced Zones

1. Human subjects, who carried out such exchanges, learned about the material world through embodied interactions that produced and measured natural phenomena. This historiographic style relies on the philosophical premise that all knowledge is the outcome of the use—as tools or instruments—of objects we find around us. And the language we use to describe these works seems to be continuous with our practical activities and styles of scientific reasoning. Hacking writes about a "style of scientific reasoning" that occurs in public as well as in private, by thinking, talking, arguing, and showing. He believes that scientific styles are not objective, but "have settled what it is to be objective (truths of certain sorts are what we obtain by con-

ducting certain sorts of investigations, answering certain standards)." Ian Hacking, "Style for Historians and Philosophers," in his *Historical Ontology* (Cambridge, MA: Harvard University Press, 2002), 180–81.

2. Eisenstein's work on print has been pioneering, stressing that it was the print culture that gave texts *standardization*, *dissemination*, and *fixity*. Hence, the possibility of reproduction and dissemination of a precise text permitted the "Scientific Revolution": research and cumulative achievement no longer had to suffer the medieval European and early modern non-European "scribe culture" and its corruption of texts. Elizabeth Eisenstein, *The Printing Revolution in Early Modern Europe* (Cambridge: Cambridge University Press, 1983). According to Eisenstein's argument, Inquisitorial censorship in Italy after the Galilean affair would have contributed to a decline of science there. Building on Eisenstein, Latour conceives of "immutable mobiles" in which the collection and deployment of durable paper entities acted as the foundation of science's success. Bruno Latour, *Science in Action* (Cambridge, MA: Harvard University Press, 1987), 52, 132–44. In reviewing Eisenstein's work, Grafton qualifies the historical break between scribal culture and print culture by asserting a high level of standardization in manuscript forms and a rigorous negotiation of readers of printed books with manuscript sources. Anthony Grafton, "The Importance of Being Printed," *Journal of Interdisciplinary History* 11, no. 2 (1980): 265–86. See also Grafton's presentation of the debate between Eisenstein and Adrian Johns in his "How Revolutionary Was the Print Revolution?" *American Historical Review* 107, no. 1 (2002): 84–87.

While Eisenstein and Latour have given agency to print technology, other historians of science, such as Johns and James Secord, give agency to humans and the way they used printed objects as tools to construct natural knowledge. This, in turn, produced deeply contextualized accounts that offer a "cultural history of print," as Johns puts it, instead of a "history of print culture." Johns emphasizes the deep, local context of print culture. Arguing against Eisenstein and Latour, he suggests that the rise of science was not in the "*fixity* of the text" but in the "credibility of the author," which reflects local cultural and social forms. Johns argues that between the fifteenth and eighteenth centuries, "print culture" suffered from the piracy of printers and scribes who copied books without authorization, a problem that would be resolved under eighteenth-century copyright laws. Adrian Johns, *The Nature of the Book: Print and Knowledge in the Making* (Chicago: University of Chicago Press, 1998), 324–79. The continuing debate has been about two approaches: Eisenstein's "history of print culture" and Johns's "cultural history of print." Instead of Eisenstein's cosmopolitan perspective on the effect of print, Johns looks at "communities of printers, booksellers, readers and censors." Adrian Johns, "How to Acknowledge a Revolution," *American Historical Review* 107, no. 1 (2002): 116. Secord shows that focusing on the multiple, parallel acts of reading may help us track how readers made meanings out of the works they encountered. James A. Secord, *Victorian Sensation: The Extraordinary Publication, Reception, and Secret Authorship of "Vestiges of the Natural History of Creation"* (Chicago: University of Chicago Press, 2000). The act of "reading" printed materials brought about changes in science and figured in various social ruptures.

Chartier suggests that reading was not strictly ephemeral, but left traces in society and culture by transforming forms of sociability, permitting new modes of thought and changing people's relationship with power. Roger Chartier, *The Order of Books* (Stanford: Stanford University Press, 1994), 1–23. Finally, McKenzie argues that "new readers make new texts and their new meanings are a function of their new forms." D. F. McKenzie, *Bibliography and the Sociology of the Text* (Cambridge: Cambridge University Press, 1999), 20.

3. Frances Yates, *Giordano Bruno and the Hermetic Tradition* (Chicago: Chicago University Press, 1963).

Primary Sources

FIRST LEVEL

The sources at the heart of this book have come down to us in the form of letters, manuscripts, and books. They are either bilingual, containing columns of Latin and Arabic or Persian, or straightforward translations of Latin works into those languages. At the center of each chapter, one such source offers a departure point for an ever-expanding investigation that slowly moves from this first item to other writings by the same author. These writings make up a "first circle" of primary sources. Beyond their content, they reveal textual peculiarities that often serve as anchors and turning points for the investigation. In deducing from these items further associations and clues for new directions, the investigation expands to other sources and documents, encompassing adjacent contexts and recovering the social and cultural networks in which our protagonists were working.

Delmedigo, Joseph Solomon. "Maamar ʿal kochav shavit." MS F 64619. Institute of Hebrew Manuscripts, Hebrew National Library, Jerusalem.

———. *Melo hofanim.* Edited by Abraham Geiger. Berlin, 1860.

———. *Mitzraf ḥokhmah.* Warsaw, 1890.

———. *Novlot ḥokhmah.* Basel, 1631.

———. *Sefer Elim.* Amsterdam, 1628. Reprint, Odessa, 1865.

Duret [Durett], Noël. *Nouvelle théorie des planètes: Conforme aux obseruations de Ptolomée, Copernie, Tycho, Lansberge, & autres excellens Astronomes, tant anciens que modernes. Avec les tables Richeliennes et Parisiennes, exactement calculées.* Paris, 1635.

———. *Novæ motuum cælestium ephemerides Richelianæ annorum 15, ab anno 1637 incipientes, ubi sex anni priores e fonti bus Lansbergianis, reliqui vero e numeris Tychoni-Keplerianis eruntur, quibus accesserunt. In priori parte. 1 Isagoge in Astrologiam. 2 De aeris mutatione. 3 Doctrina primi mobilis exquisite demonstrata. In secunda parte. 1 Usus tabularum astronomicarum pro rebus omnibus ad astronomiam spectantibus instituendis. 2 De crisium mysterio tractatus. 3 Gnomonices liber unus . . .* London, 1647.

———. *Primi mobilis doctrina dvabvs partibvs contenta. In priori qvidem rectas circvlo adscriptas, circulorum ac triangulorum affectiones cum exquisitis vbique demonstrationibus complectitur. In posteriori verè eorumdem analysin vsumque continet.* Paris, 1637.

———. *Supplementi tabularum Richelienarum, pars prima, cum brevi planetarum, Theoria ex Kepleri, sententia.* Paris, 1639.

Greaves, John. "An Account of the Latitude of Constantinople, and Rhodes, Written by the Learned Mr. John Greaves, Sometime Professor of Astronomy in the University of Oxford, and Directed to the Most Reverend James Ussher Archbishop of Armagh." *Philosophical Transactions of the Royal Society of London* 15 (1685): 1295–1300. Reprint, New York: Johnson Reprint Corp.; Kraus Reprint Corp., 1963.

———. *Astronomica quaedam ex traditione Shah Cholgii Persae: una cum hypothesibus planetarum: item excreta quaedam ex Alfergani elementis astronomicis, et Ali Kustigii de terrae magnitudine et sphaerarum coelestium a terra distantiis: cum interpretatione Latina.* London, 1652.

———. *Binae tabulae geographicae une Nassir Eddini Persae, altera Vlug Beigi Tatari.* London, 1652.

———. *A Discourse of the Roman Foot and Denarius.* In *Miscellaneous Works of Mr. John Greaves Professor of Astronomy in the University of Oxford*, vol. 1. London, 1737.

———. *Elementa linguae Persicae authore Johanne Gravio: Item Anonymus Persa de sigils Arabum & Persarum astronomicis.* London, 1649.

———. *Epochæ celebriores, astronomicis, chronologicis, chataiorum, Syro-Græcorum, Arabum, Persarum, chorasmiorum, usitatæ, ex traditione Ulug Beigi, Indiæ cirta extraq; gangem principis, eas primus publicavit, recensuit, & commentariis illustravit, Johannes Gravius.* London, 1650.

———. *Miscellaneous Works of Mr. John Greaves Professor of Astronomy in the University of Oxford.* Edited by Thomas Birch. 2 vols. London, 1737.

———. *Pyramidographia; Or, A Description of the Pyramids in Ægypt.* In *Miscellaneous Works of Mr. John Greaves Professor of Astronomy in the University of Oxford*, vol. 1. London, 1737.

———."Reflexions on a Report Made by Lord Treasurer Burleigh to the Lords of the Council . . . as Concerning the Needful Reformation of the Vulgar Kalendar, for the Civil Years and Days Accompting of Verifying According to the Time Truly Spent." *Philosophical Transactions of the Royal Society of London* 21 (1699): 343–54.

Ibn-Ma'ārūf, Taqī al-Dīn Muḥammad. "Baqiyyat al-ṭullāb ilā 'ilm al-ḥisāb" (1578). Topkapi Palace Museum Library, Istanbul.

———. *Kitāb al-kawākib al-durrīyah fī bānkāmāt al-dawrīyah.* Mīqāt Collection, MS 557/1. Dār al-Kuttub, Cairo.

———. *Kitāb al-ṭuruq al-samiyyah fī al-ālāt al-rūḥānīyyah.* Facsimile in *Taqī al-Dīn wa al-handasah al-mīkānīkiyyah*, edited by Aḥmad Yūsuf al-Ḥasan. Aleppo: University of Aleppo, 1976.

———. *Sidrat al-muntah al-afkār fi malkūt al-falak al-dawār al-zīj al- Shāhinshāhī.* MS 2930. Nuruosmaniye Library, Süleymaniye Kütüphanesi, Istanbul.

al-Manṣūr, 'Alā' al-Dīn. *Shāhinshāhnāma.* MS F 1404. Istanbul University Library, Istanbul.

Valle, Pietro della. *De viaggi di Pietro della Valle il pellegrino: descritti da lui medesimo in lettere familiari all'erudito suo amico Mario Schipano.* 2 vols. Rome, 1650.

———. "Letter to Al-dīn al-Lārī." Persian Collection, MS 9. Vatican Library, Vatican City.

al-Zigetvari Tezkireci, Ibrāhīm Efendi. *Sajanjal al-aflāk fi ghāyat al-idrāk* (1664). Kandilli Rasthanesi Collection, MS 403. Istanbul.

Zübdet al-Tavarih (Istanbul, 1580). Chester Beatty Library, Dublin.

SECOND LEVEL

In a "second circle" of sources, we find materials used by our protagonists, materials that were produced by their colleagues, patrons, and even their intellectual and political opponents. Reading the first and second circles of sources against each other supplies corroborations and clues that help answer questions emerging from reading the first circle of primary sources.

Abū-Ma'shar. *Kitāb al-milal wa al-duwal.* Edited by K. Yamamoto and C. Burnett. 2 vols. Leiden: Brill, 2000.

Africanus, Leo. *A History and Description of Africa.* Translated by John Pory (1600). 3 vols. London, 1896.

Aldimari, B. *Historia genealogica della familiglia Carafa.* Naples, 1691.

Almosnino, Rabbi Moshe. "Sefer haSefira." Translation from the 1560s by George Peurbach. In *Theoricae novae planetarum.* MS F 65321. Institute of Hebrew Manuscripts, National Library of Jerusalem, Jerusalem.

al-Baghdadi, Muhammad. *De superficierum divisionibus liber; Federici Commandini de eadem re libellus.* Pesaro, 1570.

Bainbridge, John. *An Astronomical Description of the Comet of 1618.* London, 1619.

———. *Canicularia.* London, 1652.

———. *Procli sphaera et Ptolmaei de hepothesibus planetarum.* London, 1620.

Baldi, Bernardino. *Vita di Federico Commandino* (1587). In *Giornale de' Letterati d'Italia*, vol. 19. Venice, 1714.

———. *Le vite de' matematici: edizione annotata e commentata della parte medievale e* rinascimentale. Milan: F. Angeli, 1998.

ben Elijah, Aaron. *Etz hayyim.* Leipzig, 1841.

Ben ha-Kanah, Nehunya. *Sefer haBahir.* Jerusalem: Baḳal 1974.

Ben-Israel, Menasseh. *The Hope of Israel: The English Translation of Moses Wall, 1652; Edited, with Introduction and Notes by Henry Méchoulan and Gérard Nahon; Introduction and Notes Translated from the French by Richenda George.* London, 1650.

Ben Moses, Israel, Cardinal Richelieu, and Voisin de Joseph. *Dispvtatio Cabalistica R. Israel Filii R. Mosis De Anima Et Opus Rhythmicum R. Abraham Abben Ezræ de modis, quibus Hebræilegem solent interpretari: Verbum de verbo expressum extulit Nobilis Ioseph De Voysin. Adiectis commentariis ex Zohar.* Paris, 1635.

Ben-Shealtiel, Zerubavel. *Hazon Zerubavel.* Istanbul, 1503.

Birch, Thomas, ed. *Miscellaneous Works of Mr. John Greaves Professor of Astronomy in the University of Oxford*. 2 vols. London, 1737.

Borri, Cristoforo [Christopher Borrus]. *An Account of Cochin-China: The First Treatise of the Temporal State of That Kingdom and Second of What Concerns the Spiritual* (1633). London, 1704.

———. *Collecta astronomica: ex doctrina P. Christophori Borri, mediolanensis, ex Societate Iesu; Detribuscaelis. Aereo, Sydereo, Empyreo; Issu, et studio Domini D. Gregorii de Castelbranco Comitis Villae Nouae, Sorteliae, & Goesiae domus dynastae, Regij corporis Cnstodi maximo, &c. Opus sane mathematicum, philosophicum, & theologicum, sive scrpturarium. Superiorum permissu*. Lisbon, 1631.

Brahe, Tycho. *Tycho Brahe's Description of His Instruments and Scientific Work, as Given in Astronomiae instauratae mechanica* (1598). Translated and edited by Hans Raeder, Elis Strömgren, and Bengt Strömgren. Copenhagen: I kommission hos E. Munksgaard, 1946.

———. *Tycho Brahe His Astronomicall Conjectur of the New and Much Admired Which Appeared in the Year 1572*. London, 1632.

———. *Tychonis Brahe Dani: opera omnia*. Edited by J. L. E. Dreyer. 15 vols. Copenhagen: Gyldendal, 1913–29.

Brerewood, Edward. *Enquiries touching the Diversities of Languages and Religions*. London, 1674.

Breves [Brev], François Savary de. *Seigneur François Savary, Articles du traité fait entre Henri le Grand, roi de France & de Navarre; & Sultan Amat, empereur des Turcs, en l'année 1604: Par l'entremise de messire François Savary, seigneur de Breves, conseiller du roi en ses conseils d'etat & privé, lors ambassadeur pour sa majesté à la porte dudit empereur. L'empereur Amat fils de l'impereur Mehemet toûjours victorieux*. Paris, 1604.

Bricot, Thomas. *Toldot HaAdam*. Translated by David Ben-Shushan. MS 5475. Jewish Theological Seminary, New York.

Briggs, Henry, and Henry Gellibrand. *Trigonometria Britanica*. London, 1658.

Bruno, Cristóvão. *Arte de navegar* (1628). Lisbon: República Portuguesa, 1940.

Budovec, Václav. *Anti-Alkoran, to jest mocní a nepřemožení důvodové toho, ze Al-Koran Turecký z d'ábla pošel*. Prague, 1614.

Bukhārī, Muḥammad Ibn Ismāʿīl. *Kitāb al-fitan wa ashrāṭ al-sāʿah*. In *Ashrāṭ al-sāʿah fī mustanad al-imām Aḥmad wa-zawāʾid al-ṣaḥīḥayn: jamaʿan wa-takhrījan wa-sharḥan wa-dirasah*. Edited by Khālid ibn Nāṣir ibn Saʿīd Ghāmidī. Jiddah: Dār al-Andalus al-Khaḍrāʾ, 1999.

Campanella, Tommaso. *Apologia per Galileo: testo Latino a fronte, acura di Paolo Ponzio*. Milan: Rusconi, 1997.

———. *Apologiae pro Galileo: A Defense of Galileo the Mathematician from Florance, Which Is an Inquiry as to Whether the Philosophical View Advocated by Galileo Is in Agreement with, or Is Opposed to, the Sacred Scriptures*. Translated by Richard J. Blackwell. Notre Dame, IN: University of Notre Dame Press, 1994.

———. "The Defense of Galileo." Translated by Grant McColley. *Smith College Studies in History* 22, no. 3–4 (1937).

Christie's Gallery. *The Murad III Globes: The Property of a Lady to Be Offered as Lot 139 in a Sale of Valuable Travel and Natural History Books, Atlases, Maps, and Important Globes on Wednesday 30 October 1991* . . . London: Christie, Manson & Woods, 1991.

Collet, John. *A Treatise of the Future Restoration of the Jews and Israelites to Their Land: With Some Account of the Goodness of the Country, and Their Happy Condition There, Till They Shall Be Invaded by the Turks: With Their Deliverance from All Their Enemies, When the Messiah Will Establish His Kingdom at Jerusalem, and Bring in the Last Glorious Ages*. London, 1774.

Conforte, David. *Kore haDorot*. Berlin, 1846.

Copernicus, Nicolaus. *Complete Works* (in English). 4 vols. London: Macmillan, 1972–1992.

Cordovero, Moses. *Sefer Elima*. Jerusalem: Or ḥadash, 1998.

Culpeper, Nicholas. *Semiotica urinica: Astrological Judgment of Diseases*. 4th ed. London, 1671.

Daniel, Lundius. *Commentarius R. Aben Esrae in Prophetam Habacuc: Quem ex Hebraeo in Latinum sermonem versum & brevibus notis illustratum*. Upsal, 1706.

Dury, John. "An Epistolicall Discours, to Mr. Thorowgood, Concerning his Conjecture That the Americans Are Decended from the Israelites." In *Jews in America, or Probabilities That the Americans Are of That Race*, edited by Thomas Thorowgood. London, 1650.

Flamsteed, John. *The Correspondence of John Flamsteed, the First Astronomer Royal*. Bristol and Philadelphia: Institute of Physics Publishing, 2002.

Foscarini, Paolo Antonio. *Concerning the Pythagorian and Copernican Opinion of the Mobility of the Earth and Stability of the Sun and of the New Systeme or Constitution of the World*. In *Mathematical Collections and Translations*, edited by Thomas Salusbury. London, 1667.

Gaffarel, Jacques. *Unheard-of Curiosities: Concerning the Talismanical Sculpture of the Persians; the Horoscope of the Patriarkes; and the Reading of the Stars*. Translated by Edmund Chilmead. London, 1650.

Galileo, Galilei. *Le opere di Galileo Galilei*. 20 vols. Florence, 1890–1909.

Gassendi, Pierre. *Tychonis Brahei, equitis Dani, astronomorum coryphaei, vita*. Paris, 1654.

Ghāyat al-ḥakīm wa-aḥaqq al-natījatayn bi-aītaqdīm = Picatrix, Das Ziel des Weisen. Berlin: B. G. Teubner, 1933.

Giannone, Pietro. *The Civil History of the Kingdom of Naples*. Translated by James Ogilvie. 2 vols. London, 1723.

Giustiniani, Agostino. *Psalterim Hebreum, Grecum, Arabicum, et Chaldeum cum tribus Latinis interpretationibus et glosis . . . Aug. Luistinaini genuensis praedictorii ordinis episcopi calendis Augustii*. Genoa, 1516.

Godinho, Manuel. *Intrepid Itinerant: Manuel Godinho and His Journey from India to Portugal in 1663*. Edited by John Correia-Afonso. Translated by Vitalio Lobo and John Correia-Afonso. Bombay: Oxford University Press, 1990.

HaLevi, Abraham Ben-Eliezer. *Ma'amar Meshare Qitrin* (Constantinople, 1510). Edited by Gershom G. Scholem. Jerusalem: Jewish National & University Library Press, 1977.

Ibn ʿAbd al-Ghanī, Aḥmad Shalabī. *Awḍah al-ishārāt fīman tawallá Miṣr al-Qāhirah min al-wuzarāʾ wa-al-bāshāt: al-mulaqqab bi-al-tārīkh al-ʿaynī*. Cairo: Tawzīʿ Maktabat al-Khānjī, 1978.

Ibn al-Haytham, al-Ḥassan Ibn Ḥusayn. *al-Shukūk ʿalā Baṭlamyūs* (*Dubitationes in Ptolemaeum; Aporias against Ptolemy*). Edited by A. I. Ṣabra and Nabīl Shihābī. Cairo: Maṭbaʿat Dār al-Kutub, 1971.

Ibn Kathīr, Ismāʿīl ibn ʿUmar. *Kitāb al-nihāyah, aw al-fitan wa-al-malāḥim*. Edited by Ṭāhā Muḥammad al-Zaynī. Cairo: Dār al-Kutub al-Ḥadīthah, 1969.

al-Kāshī, Ghiyāth al-Dīn Jamshīd. *Uluğ Bey Ve Semarkanddeki Ilim Faaliyeti Hakkindu Giyasüddin Kâşi'nin Mektubu = Ghiyâth al Dîn al Kâshî's letter on Ulugh Bey and the Scientific Activity in Samarqand*. Edited by Aydin Sayili. Ankara: Türk Tarih Kurumu Basımevi, 1960.

Kepler, Johannes. *Gesammelte Werke*. Edited by Max Caspar. Munich: Beck, 1954.

———. *Tabulæ Rudolphinæ; Or, The Rudolphine Tables: Supputated to the Meridian of Uraniburge, First by John Kepler; from the Observations of the tres noble Ticho Brahe, afterwards Digested into a Most Accurate, and Easie Compendium, by the Famous Johannes Baptista Morinvs and Printed for Him at Paris, Anno Dom. 1650*. London, 1675.

La Lande, Jèrôme de. *Bibiographie astronomique: avec l'histoire de l'astronomie depuis 1781 jusquà 1802*. Paris, 1803.

Laud, William. *The Works of the Most Reverend Father in God, William Laud, D.D., Sometime Lord Archbishop of Canterbury*. 7 vols. Oxford, 1847.

Luzzatto, Simone. *Discorso circa il stato degli Hebrei*. Venice, 1638.

———. *Socrate overo dell'humano sapere*. Venice, 1651.

Modena, Leon da [Judah Arieh of Modena]. *Behinat haCabala*. Jerusalem: Sifriyah le-maḥshevet Yiśraʾel, 1968.

———. *The History of the Rites, Customes, and Manner of Life, of the Present Jews, throughout the World. Written in Italian, by Leo Modena . . . Tr. into English, by Edmund Chilmead*. London, 1650.

———. *Letters of Rabbi Leon Modena*. Edited by Yacob Boksenboim. Tel Aviv: Bet hasefer le-madaʿe ha-Yahadut ʿal shem Ḥayim Rozenberg, Universiṭat Tel Aviv, 1984.

"Mr. Flamsteads Letter of July 24 1675 to the Publisher, relating to Another, Printed in Num. 110 of These Tracts, concerning M. Horroxes Lunar System." *Philosophical Transactions of the Royal Society of London* 10 (1675).

Newton, Isaac. *A Dissertation upon the Sacred Cubit of the Jews and the Cubits of the Several Nations in Which, from the Dimensions of the Greatest Egyptian Pyramid, as Taken by Mr. John Greaves, the Antient Cubit of Memphis Is Determined*. In *Miscellaneous Works of Mr. John Greaves Professor of Astronomy in the University of Oxford*, edited by Thomas Birch, vol. 2. London, 1737.

Parr, Ri. *A Collection of Three Hundred Letters: Written . . . James Usher, Late Lord Archbishop of Armagh . . . and others; as . . . Is. Vossius, Hu. Grotius a.o.* London, 1686.

Pocock, Edward. *Bar Hebraeus, Specimen historiae Arabum*. Oxford, 1650.

——. *Porta Mosis*. Oxford, 1655.

Popkin, Richard. "Les Caraites et l'emancipation des Juifs." *Dix-Huitième Siècle* 13 (1981): 137–47.

——. "The Lost Tribes, the Caraites and the English Millenarians." *Journal of Jewish Studies* 37, no. 2 (1986): 213–27.

Predictions of the Sudden and Total Destruction of the Turkish Empire and Religion of Mahomet: According to the Opinions of the Lord Tycho Brahe of Denmark, and Many Others of the Best Astronomers of This Later Age, Collected and Humbly Dedicated to all Christendom by a Lover of Christianity. London, 1684.

Reymers, Nicolas [Raimarus Ursus]. *Chronological, Certain, and Irrefutable Proof, from the Holy Scripture and Fathers; That the World Will Perish and the Last Days Will Come within 77 Years*. Nuremberg, 1606.

Rittangel, Johann Stephan. *Sefer Yetsirah; id est, liber Iezirah qui Abrahamo Patriarchæ adscribitur, unà cum commentario rabi Abraham f. D super 32 semitis sapientiæ, à quibus liber Iezirah incipit*. Amsterdam, 1642.

Salusbury, Thomas. *Mathematical Collections and Translations in Two Parts from the Original Copies of Galileus, and Other Famous Modern Authors*. London, 1667.

Sambari, Yosef ben Yitzhak. *Sefer Divrei Yosef*. Edited by Shimon Shtober. Jerusalem: Yad Yitsḥaḳ Ben-Tsevi ṿeha-Universiṭah ha-ʻIvrit bi-Yerushalayim, 1994.

Schweigger, Salomon. *Ein newe Reyssbeschreibung auss Teutschland nach Constantinopel und Jerusalem*. Edited by Rudolf Neck Wein. Graz: Akademische Druck- u. Verlagsanstalt, 1964.

——. *Der Turken Alcoran, Religion und Aberglauben*. Nuremberg, 1616, 1623.

Soranzo, Lazaro. *L'Othomanno*. Ferrara, 1598.

Suhrawardi, Shihāb al-dīn. *Ḥikmat al-ishrāq*. Translated and edited by John Walbridge and Hossein Ziai. Salt Lake City, UT: Brigham Young University Press, 1999.

Thomas, Herbert. *Some Years Travels into Divers Parts of Africa and Asia the Great Describing More Particularly the Empires of Persia and Industan*. London, 1662.

Tishbi, Yoseph. *Hezionot vezihronot* (Constantinople, 1579). F 21357. Institute of Hebrew Manuscripts, Jerusalem.

al-Ṭūsī, Naṣīr al-Dīn. *Contemplation and Action: The Spiritual Autobiography of a Muslim Scholar [Sayr wa sulūk]*. Edited and translated by S. J. Badakhchani. London: I. B. Tauris, in association with Institute of Ismaili Studies, 1998.

——. *al-Murāsalāt bayna Ṣadr al-Dīn al-Qaūnawī wa-Naṣīr al-Dīn al-Ṭūsī: taḥqīq wa-mulakhkhas al-Almāni mufassar Kudrun Shubart*. Beirut: Yuṭlabu min Dār al-Nashr Frānts Shtāynar, 1995.

——. *Zubdat al-idrāk fī hay'at al-aflāk: ma'a dirāsat al-manhaj al-Ṭūsī al-'ilmī fī majāl al-falak*. Edited by 'Abbās Muḥammad Muḥammad Ḥasan Sulaymān. Alexandria: Dār al-Ma'rifah al-Jāmi'īyah, 1994.

Ussher, James. *The Causes of the Continuance of the Contentions concerning the Church Government*. In *The Whole Works of the Most Rev. James Ussher*, edited by Charles Richard Elrington, vol. 3. Dublin, 1864.

——. *De Macedonum et Asianorum anno solari*. In *The Whole Works of the Most Rev. James Ussher*, edited by Charles Richard Elrington, vol. 7. Dublin, 1864.

———. *Geographical and Historical Disquisition touching the Asia Properly So Called.* In *The Whole Works of the Most Rev. James Ussher*, edited by Charles Richard Elrington, vol. 7. Dublin, 1864.

———. *Gravissimae quaestionis de Christianrum Ecclesiarum* . . . Dublin, 1864.

———. *Tractatus de controversiis pontificiis.* In *The Whole Works of the Most Rev. James Ussher*, edited by Charles Richard Elrington, vol. 14. Dublin, 1864.

Wilkins, John. *A Discourse Concerning a New World & Another Planet.* London, 1640.

———.*The Discovery of a World in the Moone; Or, Discourse Tending to Prove That 'Tis Probable There May Be Another Habitable World in That Planet.* London, 1638.

———. *An Essay towards a Real Character and a Philosophical Language.* London, 1668.

Reference Works

Biographical dictionaries, catalogues, and chronicles of the early modern period and later enable one to recover the social and cultural contexts in which a work was written.

Accattatis, Luigi. *Le biografie degli uomini illustri delle Calabrie.* 2 vols. Cosenza, 1869.

Bartoli, D. *Della vita del P. Vincenzo Carafa, settimo generale della compagnia di Giesú.* Rome, 1651.

Bayle, Pierre. *Dictionnaire historique et critique: The Dictionary Historical and Critical of Mr. Peter Bayle.* Translated into English from the third edition of the original French. 5 vols. London, 1734–38.

Buzurg, Agha. *al-Dharī'ah ilá taṣānīf al-Shī'ah, ta'līf Muḥammad al-shahīr bi-al-Shaykh Āghā Buzurg al-Ṭihrānī.* al-Najaf: Maṭba'at al-Gharrī, 1936.

———. *Tabaqat a'lam al-Shi'ah.* Beirut, 1971.

Çelebī, Kātip [Kātib]. *The Balance of Truth (Mīzān al-ḥaqq).* Translated by G. L. Lewis. London: Allen and Unwin, 1957.

———. *Kashf al-zunūn.* 2 vols. Istanbul, 1943.

———. *Sullam al-ūsūl ilā tabakāt al-fuhūl.* Shehid Ali Pasha Section, MS 1887. Süleymaniye Library, Istanbul.

Habermann, Abraham Meir. *Ha-Madpis Daniyel Bombirg u-reshimat sifre bet defuso.* Tsefat: ha-Muse'on le-'omanut ha-defus, 1978.

———. *Ha-Madpiss Cornelio Adel Kind u-beno Daniel: u-reshimath haSefarim sheNidpessu al yedehem.* Jerusalem: R. Mas, 1980.

Ibn al-Zubayr, Muḥammad, and al-Sa'īd Muḥammad Badawī. *Sijill asmā' al-'Arab* Beirut: Maktabat Lubnān, 1991.

İhsanoğlu, Ekmeleddin, ed. *Bibliography on Manuscript Libraries in Turkey and the Publications on the Manuscripts Located in These Libraries.* Istanbul, 1995.

———. *Büyük Cihad'dan Frenk fodulluğuna.* Istanbul: İletişim, 1996.

———. "Introduction." In *Catalogue of Manuscripts in the Köprülü Library*, compiled by Cemil Akpinar, Cevat Izgi, and Ramazan Sesen. 3 vols. Istanbul: Research Centre for Islamic History, Art, and Culture 1986.

———. "Introduction of Western Science to the Ottoman World: A Case Study of Modern Astronomy (1660–1860)." In *The Transfer of Modern Science and Technology to the Muslim World*, edited by Ekmeleddin İhsanoğlu. Istanbul: Research Centre for Islamic History, Art, and Culture, 1992.

———, ed. *Osmanlı astronomi literatürü tarihi.* 2 vols. Istanbul: Research Centre for Islamic History, Art, and Culture, 1997

———, ed. *Union Catalogue of Periodicals in Arabic Script.* Istanbul: Research Centre for Islamic History, Art, and Culture, 1986.

Jihāmī, Jīrār. *Mawsūʿat muṣṭalaḥāt al-ʿulūm ʿinda al-ʿArab.* 2 vols. Beirut: Maktabat Lubnān Nāshirūn, 1999.

King, David, ed. *Fihris al-makhṭūṭāt al-ʿilmīyah al-maḥfūẓah bi-Dār al-Kutub al-Miṣrīyah.* Cairo: al-Hayʾah al-Miṣrīyah al-ʿĀmmah lil-Kitāb: Bi-al-taʿāwun maʿa Markaz al-Buḥūth al-Amrīkī bi-Miṣr wa-Muʾassasat Smīthsūnyān, 1981.

Lenzi, Luigi Aliquò, and Filippo Aliquò Taverriti. *Gli scrittori Calabresi: dizionario bio-bibliografico.* 4 vols. Reggio di Calabria: Corriere di Reggio, 1955.

Munzavī, Aḥmad. *Fihristvārah-i kitābhā-yi Fārsī.* Tehran: Anjuman-i Āṣār va Mafākhir-i Farhangī, 1995.

Riḍa, Kaḥḥāla ʿUmar. *Muʿjam al-muʾallifīn: tarājim muṣannifī al-kuttub al-ʿArabiyyah.* 9 vols. Beirut: Dār Iḥyāʾ al-Turāth al-ʿArabī, 1957.

Steinschneider, Maurice. *Ozar ha-sepharim: Tesaurus liberorum hebraicorum tam impressorum quam many scriptorum.* Vilnius, 1880.

Tabīb, Rashīd al-Din. *Rashiduddin Fazlullah's Jamiʿu't-tawarikh = Compendium of Chronicles: A History of the Mongols.* Translated and annotated by W. M. Thackston. 3 vols. Cambridge, MA: Harvard University, Department of Near Eastern Languages and Civilizations, 1999.

Valente, Gustavo. *Calabria Calabresi e Turcheschi nei secoli della pirateria (1400–1800).* Frama: Chiaravelle Centrale, 1973.

Venn, John, and J. H. Venn. *Alumni cantabrigienses: A Biographical List of All Known Students, Graduates, and Holders of Office at the University of Cambridge, from the Earliest Times to 1900.* Pt. 1, vol. 2. Cambridge, 1622.

Ward, John. *The Lives of the Professors of Gresham College.* London, 1740.

Wood, Anthony. *Athenæ Oxonienses . . .* Oxford, 1848.

———. *The History and Antiquities of the University of Oxford.* 2 vols. Oxford, 1792.

Yudlov, Yzhak, ed. *Giouvanni Di Gara: Printer, Venice 1564–1610, List of Books Printed at His Press.* Jerusalem: Mekhon Haberman le-meḥḳere sifrut, 1982.

Secondary Sources

Nineteenth- and twentieth-century scholarship supplies the frame of reference for early modern history of science, but also highlights the lacunas and the extent to which European and Near Eastern historiographies have become entrenched in their own cultural and paradigmatic boundaries.

Alemanno, Laura. "L'Accademia degli umoristi." *Roma moderna e contemporanea* 3, no. 1 (1995): 97–120.

Amabile, Luigi. *Fra Tommaso Campanella, la sua congiura, i suoi processi e la sua pazzia*. 3 vols. Naples, 1882.

———. *Fra Tommaso Campanella ne' castelli di Napoli, in Roma e in Parigi*. 2 vols. Naples, 1887.

Apt, Adam Jared. "The Reception of Kepler's Astronomy in England: 1596–1650." Ph.D. diss., S.I., St. Catherine's College, 1982.

'Aṭā Allāh, Samīr. *Tārīkh wa-fann ṣinā'āt al-kitāb*. Beirut: Dār 'Atā Allah (Nawfal), 1993.

Attias, Jean-Christophe. *Le Commentaire biblique: Mordechai Komtiano ou l'herméneutique du dialogue*. Paris: Cerf, 1991.

Babinger, F. "Maometto II conquistatore e l'Italia." *Rivista storica italiana* 63 (1951): 469–505.

Baroncelli, Giovanna. "L'astronomia a Napoli al tempo di Galileo." In *Galileo e Napoli*, edited by Fabrizio Lomonaco and Maurizio Torrini. Naples: Guida, 1987.

Barone, Francesco. "Deigo De Zuñiga e Gelileo Galilei: Astronomia eliostatica ed esegesi biblica." *Critica storia* 3 (1982): 319–34.

Barzilay, Isaac. *Yoseph Shlomo Delmedigo (Yashar of Candia)*. Leiden: Brill, 1974.

Ben-Zaken, Avner. "Political Economy and Scientific Activity in the Ottoman Empire." In *The Turks*, edited by H. C. Güzel, C. C. Oğuz, and O. Karatay, vol. 3. Ankara: Yeni Türkiye, 2002.

———. "Recent Currents in the Study of Ottoman-Egypt Historiography, with Remarks about the Role of the History of Natural Philosophy and Science." *Journal of Semitic Studies* 49, no. 2 (2004): 303–28.

Biagioli, Mario. *Galileo, Courtier: The Practice of Science in the Culture of Absolutism*. Chicago: University of Chicago Press, 1993.

———. "Knowledge, Freedom, and Brotherly Love: Homosociality and the Accademia dei Lincei." *Configurations* 2 (1995): 139–66.

Boaga, Emanuele. "Annotazioni e documenti: sulla vita e sulle opera di Paolo Antonio Foscarini teologo 'Copernicano (1562c.–1616).'" *Carmelus* 37 (1990): 173–216.

Bono, Salvatore. *Corsari nel Mediterraneo: cristiani e musulmani fra guerra, schiavitù e commercio*. Milan: A. Mondadori, 1993.

Brotóns, Victor Navarro. "The Reception of Copernicus in Sixteenth-Century Spain." *Isis* 86, no. 1 (1995): 52–78.

Brotton, Jerry. "Printing the World." In *Books and the Sciences in History*, edited by Nicholas Jardine and Marina Frasca-Spada. Cambridge: Cambridge University Press, 2000.

Carnell, Elizabeth. "Talismans and the Stars: Varieties of Interpersonal Magic in Ghayat al-Hakim." Ph.D. diss., University of Michigan, 2002.

Caroti, Stefano. "Un sostenitore napoletano della mobilità della terra: il padre Paolo Antonio Foscarini." In *Galileo e Napoli*, edited by Fabrizio Lomonaco and Maurizio Torrini. Naples: Università degli Studi di Napoli, 1987.

Carter, Thomas. "Islam a Barrier to Printing." *Moslem World* 3 (1943): 213–16.

Cheikho, L. "Tārīkh fann al-ṭibā'a fi al-mashriq." *al-Mashriq* 3–5 (1900–1902).

Christianson, John. *On Tycho's Island: Tycho Brahe and His Assistants, 1570–1601.* Cambridge: Cambridge University Press, 2000.

———. "Tycho Brahe's Cosmology from the *Astrologia.*" *Isis* 59, no. 3 (1968): 312–18.

Clifford, James. *The Predicament of Culture: Twentieth-Century Ethnography, Literature, and Art.* Cambridge, MA: Harvard University Press, 1988.

———. "Traveling Cultures." In *Cultural Studies*, edited by Lawrence Grossberg, Cary Nelson, and Paula A. Treicher. New York: Routledge, 1992.

Copenhaver, Brian. "Natural Magic, Hermetism, and Occultism in Early Modern Science." In *Reappraisals of the Scientific Revolution*, edited by David C. Lindberg and Robert S. Westman. Cambridge: Cambridge University Press, 1990.

Dannenfeld, Karl. "The Humanists' Knowledge of Arabic." *Studies in the Renaissance* 2 (1955): 96–117.

Dávid, Gèza. "An Ottoman Military Career on the Hungarian Borders: Kasim *Voyvoda*, Bey and Pasha." In *Ottomans, Hungarians, and Habsburgs in Central Europe: The Military Confines in the Era of Ottoman Conquest*, edited by Gèza Dávid and Pál Fodor. Leiden: Brill, 2000.

Davis, Natalie Zemon. "Beyond the Market: Books as Gifts in Sixteenth-Century France." *Transactions of the Royal Historical Society*, ser. 5, 33 (1983): 69–88.

———. *The Gift in Sixteenth-Century France.* Madison: University of Wisconsin Press, 2000.

———. *Trickster Travels: A Sixteenth-Century Muslim between Worlds.* New York: Hill and Wang, 2006.

Dizer, Muammer. *Takiyüddin.* Ankara: Kültür Bakanlığı, 1990.

Dreyer, J. L. E. *Tycho Brahe: A Picture of Scientific Life and Work in the Sixteenth Century.* New York: Dover, 1963.

Evans, R. J. W. "Bohemia: The Emperor, and the Porte, 1550–1600." *Oxford Slavonic Papers* 3 (1970): 85–106.

———. *Rudolf II and His World: A Study in Intellectual History, 1576–1612.* New York: Oxford University Press, 1973.

Fleischer, Cornell. *Bureaucrat and Intellectual in the Ottoman Empire: The Historian Mustafa 'Ali.* Princeton, NJ: Princeton University Press, 1986.

———. "The Lawgiver as Messiah: The Making of the Imperial Image in the Reign of Suleyman." In *Soliman Le Magnifique et son temps*, edited by Gilles Veinstein. Paris: Documentation française, 1999.

———. *Muṣṭafā 'Alī's Description of Cairo of 1599.* Vienna: Verlag der Österreichischen Akademie der Weissenschaften, 1975.

Forster, Charles Thornton, and F. H. Blackburne Daniell, eds. *The Life and Letters of Ogier Ghiselin De Busbecq: Seigneur of Bousbeque Knight, Imperial Ambassador.* London, 1881.

Gabrieli, Giuseppe. *Contributi alla storia dell'Accademia dei Lincei.* 2 vols. Rome: Accademia nazionale dei Lincei, 1989.

———. *I primi Accademici Lincei e gli studi orientali.* Florence: Olschki, 1926.

Gèza, Dávid, and Fodor Pál, eds. *Ottomans, Hungarians, and Habsburgs in Central Europe: The Military Confines in the Era of Ottoman Conquest.* Leiden: Brill, 2000.

Gingerich, Owen. *An Annotated Census of Copernicus' De Revolutionibus (Nuremberg, 1543, and Basel, 1566).* Leiden: Brill, 2002.

Grafton, Anthony. "How Revolutionary Was the Print Revolution?" *American Historical Review* 107, no. 1 (2002): 84–87.

———. "The Importance of Being Printed." *Journal of Interdisciplinary History* 11, no. 2 (1980): 265–86.

Habib, S. Ifran, and Dhruv Raina, eds. *Situating the History of Science: Dialogues with Joseph Needham.* Oxford: Oxford University Press, 1999.

Hashimoto, Keizo. *Hsü Kuang-Ch'i and Astronomical Reform: The Process of the Chinese Acceptance of Western Astronomy, 1629–1635.* Osaka: Kansai University Press, 1988.

Haynes, Jonathan. *The Humanist as a Traveler.* London: Associated University Press, 1986.

Heninger, S. K. *Touches of Sweet Harmony: Pythagorean Cosmology and Renaissance Poetics.* San Marino, CA: Huntington Library, 1974.

Hill, Donald. *Arabic Water-Clocks.* Aleppo: University of Aleppo, 1981.

Huff, Toby. *The Rise of Early Modern Science: Islam, China, and the West.* Cambridge: Cambridge University Press, 1993.

Johns, Adrian. "How to Acknowledge a Revolution." *American Historical Review* 107, no. 1 (2002): 106–25.

Kaplan, Yosef. "'Karaites' in Early Eighteenth-Century Amsterdam." In *Sceptics, Millenarians, and Jews*, edited by David Katz and Jonathan Israel. Leiden: Brill, 1990.

Kennedy, E. S. "The Exact Sciences in Timurid Iran." In *The Cambridge History of Iran: Volume Six, The Timurid and Safavid Periods*, edited by Peter Jackson and Laurence Lockhart. Cambridge: Cambridge University Press, 1986.

Kettering, Sharon. "Gift-Giving and Patronage in Early Modern France." *French History* 2 (1988): 133–51.

King, David. *Astronomy in the Service of Islam.* London: Variorum, 1993.

———. "The Astronomy of the Mamluks." *Isis* 74 (1983): 531–55.

———. *Islamic Mathematical Astronomy.* London: Variorum, 1986.

Klein-Franke, Felix. *Astromathematics in Islam.* New York: Georg Olms Verlag, 1984.

Knorr, Wilbur. *Ancient Sources of the Medieval Tradition of Mechanics: Greek, Arabic, and Latin Studies of the Balance.* Florence: Istituto e Museo di Storia della Scienza, 1982.

Koyre, Alexander. *The Astronomical Revolution.* New York: Dover, 1973.

Kuhn, Thomas. *The Copernican Revolution.* Cambridge, MA: Harvard University Press, 1956.

———. "The Relations between History and History of Science." *Dædalus: Journal of the American Academy of Arts and Sciences*, spring (1971): 271–305.

———. *The Structure of Scientific Revolutions.* 3d ed. Chicago: University of Chicago Press, 1996.

Kuntz, Marion Leathers. *Guillaume Postel: Prophet of the Restitution of All Things; His Life and Thought.* The Hague: Martinus Nijhoff Publishers, 1981.

———. "Venezia portava el fuocho in seno: Guillaume Postel before the Council of Ten in 1548; Priest Turned Prophet." *Studi Veneziani*, n.s., 33 (1997): 95–122.

Kurz, Otto. *European Clocks and Watches in the Near East*. London: Warburg Institute, 1975.

Leibowitz, J. O. *Amatus Lusitanus (1511–1568) è Salonique*. Rome: Arti grafiche e cossidente, 1970.

Levi, Joseph. "Haakdemya hayehudit lemada bitehilat hameah hasheva-'esre: hanisayyon shel Shlomo Delmedigo." In *The Eleventh International Congress for Jewish Studies*, Section B. Jerusalem, 1984.

Lindberg, David, and Ronald Numbers, eds. *God and Nature: Historical Essays on the Encounter between Christianity and Science*. Berkeley and Los Angeles: University of California Press, 1984.

Lindberg, David C., and Robert S. Westman, eds. *Reappraisals of the Scientific Revolution*. Cambridge: Cambridge University Press, 1990.

Lockhart, Laurence. "European Contacts with Persia, 1350–1736." In *The Cambridge History of Iran: Volume Six, The Timurid and Safavid Periods*, edited by Peter Jackson and Laurence Lockhart. Cambridge: Cambridge University Press, 1986.

Lomonaco, Fabrizio, and Maurizio Torrini, eds. *Galileo e Napoli*. Naples: Università degli Studi di Napoli, 1987.

Malcolm, Noel. "The Crescent and the City of the Sun: Islam and the Renaissance Utopia of Tommaso Campanella." *Proceedings of the British Academy* 125 (2005): 41–67.

Manuzio, Paolo. *Lettere di Paolo Manuzio copiate sugli autografi esistenti nella Biblioteca Asbrosiana*. Paris, 1843.

Marshall, F. H. "An Eastern Patriarch's Education in England." *Journal of Hellenic Studies* 40 (1926): 185–202.

Masraf, Najji Zayn al-Dīn. *Bada'i' al-khatt al-'Arabi*. Baghdad: Mudiriyat al-Thaqafah al-'Ammah, 1972.

Mercati, Angelo. "Notizie sul gesuita Cristoforo Borri e su sue 'Inventioni' da carte Finora sconosciute di Pietro Della Valle il pellegrino." *Pontificia academia scientiarvm acta* 15, no. 3 (1951–53): 25–46.

Mioni, Elpidio. "Bessarione e la caduta di Constantinopoli." In *Miscellanea Marciana*, edited by Marino Zorzi, vol. 6. Venice: Beblioteca Nazionale Marciana, 1991.

Mirza, R. Muhammad, and Muhammad Iqbal Siddqi. *Muslim Contribution to Science*. Lahore: Kazi, 1986.

Moesgaard, Kristian P. "Thābit Ibn-Qurrā between Ptolemy and Copernicus." In *Avet, avec, après Copernic: La reprèsentation de l'univers et ses consèquences èpistèmologiques; XXXIe semaine de synthèse, 1–7 juin 1973*. Paris: Blanchard, 1975.

Mordtmann, J. H. "Das Observatorium des Taqī en-din zu Pera." *Der Islam* 13 (1923): 82–96.

Mu'in, Muhammad. "Naṣīr al-Dīn al-Ṭūsī wa zabān wa adab pārsī." *Majalla yi Danishkada-yi adabiyyāt* 3, no. 4 (1956): 30–42.

Neal, Katherine. "Mathematics and Empire, Navigation and Exploration: Henry Briggs and the Northwest Passage Voyage of 1613." *Isis* 93, no. 3 (2002): 435–53.

Nemoy, Leon. *Karaite Anthology.* New Haven, CT: Yale University Press, 1952.
Pamuk, Orhan. *The White Castle.* New York: Braziller, 1991.
Panicati, Roberto. *Orologi e orologiai del Rinascimento italiano: la scuola urbinate = L'horlogerie d'interieur italienne au XVIe siècle et l'école du duché d'Urbino = Sixteenth century Italian chamber clocks and the Urbino school.* Urbino: Quattro Venti, 1988.
Paolino, Giangiuseppe Origlia. *Istoria dello studio di Napoli.* Naples, 1753.
Patterson, W. B. *King James VI and I and the Reunion of Christendom.* Cambridge: Cambridge University Press, 1997.
Pedersed, Johannes. *The Arabic Book.* Translated by Geoffrey French. Princeton, NJ: Princeton University Press, 1984.
Penrose, Boies. *Travel and Discovery in the Renaissance, 1420–1620.* Cambridge, MA: Harvard University Press, 1953.
Pingree, David, ed. *Picatrix: The Latin Version of the Ghāyat al-Hakim.* London: Warburg Institute, 1986.
Qadir, C. A. *Philosophy and Science in the Islamic World.* London: Routledge, 1990.
Ragep, F. J. *Naṣīr al-Dīn al-Ṭūsī's Memoir on Astronomy (al-Tadhkirat fi 'ilm al-hay'a).* 2 vols. New York: Springer-Verlag, 1993.
Ragep, F. J., and S. Ragep, eds. *Tradition, Transmission, Transformation: Proceedings of Two Conferences on Pre-Modern Science Held at the University of Oklahoma.* Leiden: Brill, 1996.
Richard, Francis. *Raphaël du Mans, missionnaire en Perse au XVIIe sicle.* Paris: Société d'histoire de l'Orient, L'Harmattan, 1995.
Robinson, Francis. "Technology and Religious Change: Islam and the Impact of Print." *Modern Asian Studies* 27, no. 1 (1993): 229–251.
Rose, Paul Lawrence. "Humanist Libraries and Renaissance Mathematics: The Italian Libraries of the Quattrocento." *Studies in the Renaissance* 20 (1973): 46–105.
———. *The Italian Renaissance of Mathematics: Studies on Humanists and Mathematicians from Petrarch to Galileo.* Geneva: Librairie Droz, 1975.
Rosenblatt, Norman. "Joseph Nasi: Court Favorite of Selim II." Ph.D. diss., University of Pennsylvania, 1957.
Rosenthal, Frank. "The Study of the Hebrew Bible in Sixteenth-Century Italy." *Studies in the Renaissance* 1 (1954): 81–89.
Rossum, Gerhard Dohrn-van. *History of the Hour: Clocks and Modern Temporal Orders.* Chicago: University of Chicago Press, 1996.
Rouillard, Clarence Dana. *The Turk in French History, Thought, and Literature (1520–1660).* Paris: Boivin, 1940.
Ruderman, David. *Jewish Thought and Scientific Discovery in Early Modern Europe.* New Haven, CT: Yale University Press, 1995.
Russel, G. A. ed. *The 'Arabick' Interest of the Natural Philosophers in Seventeenth-Century England.* Leiden: Brill, 1994.
Sabra, A. I. "Configuring the Universe: Aporetic, Problem Solving, and Kinematic Modeling as Themes of Arabic Astronomy." *Perspectives on Science* 6, no. 3 (1998): 35–36.

———. *Optics, Astronomy, and Logic Studies in Arabic Science and Philosophy*. Brookfield, VT: Variorum, 1994.

———. "Situating Arabic Science: Locality versus Essence (History of Science Distinguished Lecture)." *Isis* 87 (1996): 654–70.

Saliba, George. *A History of Arabic Astronomy: Planetary Theories during the Golden Age of Islam*. New York: New York University Press, 1994.

———. "Mediterranean Crossings: Islamic Science in Renaissance Europe." Paper presented at Stanford University. Fall 1998.

———. [Review of Huff, *The Rise of Early Modern Science*]. *Bulletin of the Royal Institute for Inter-Faith Studies (BRIIFS)* 1, no. 2 (1999).

———. "A Sixteenth-Century Arabic Critique of Ptolemaic Astronomy: The Work of Shams al-Din al-Khafri." *Journal for the History of Astronomy* 25 (1994): 15–38.

———. "Writing the History of Arabic Astronomy: Problems and Differing Perspectives." *Journal of the American Oriental Society* 116, no. 4 (1996): 709–19.

Sarkīs, Yusuf Ilyan. *Muʿjam al-maṭbūʿāt al-ʿArabīyah wa-al-muʿarrabah . . . min yawm ẓuhūr al-ṭabāʿah ilā nihāyat al-sanah al-hijriyyah 1339 al-muwāfiqāh li-sanat 1919 mīlādiyyah.* Cairo: Sarkīs, 1928.

Sayili, Aydin. "ʿAlāʾ al-Dīn al-Manṣūr's Poems on the Istanbul Observatory." *Türk Tarīh Kurumu Belleten* 79, summer (1956): 429–84.

———. "The Causes of the Decline of Scientific Work in Islam." In *The Observatory in the Islam and Its Place in the General History of the Observatory*, edited by Aydin Sayili. Ankara: Türk Tarih Kurumu Basımevi, 1960.

———, ed. *The Observatory in Islam and Its Place in the General History of the Observatory.* Ankara: Türk Tarih Kurumu Basımevi, 1960.

———. *Tycho Brahe Sistemi Hakkinda xvii. Asir Baslarina Ait Farça Bir Yazma.* Ankara: Dil ve Tarih-Cografya Fakültesi, 1958.

Schechner, Sara J. *Comets, Popular Culture, and the Birth of Modern Cosmology.* Princeton, NJ: Princeton University Press, 1997.

Schimmel, Annemarie. "Calligraphy and Sufism in Ottoman Turkey." In *The Dervish Lodge: Architecture, Art, and Sufism in Ottoman Turkey*, edited by Raymond Lifchez. Berkeley and Los Angeles: University of California Press, 1992.

———. *Mystical Dimension of Islam.* Chapel Hill: University of North Carolina Press, 1975.

Scholem, Gershom Gerhard. *Sefer haZohar shel Gershom Scholem ʾim he ʾarot bikhetivat yado*. Jerusalem: Hebrew University Press, 1982.

Setton, Kenneth M. *Western Hostility to Islam and Prophecies of Turkish Doom*. Philadelphia: American Philosophical Society, 1992.

Shalev, Zur. "Measurer of All Things: John Greaves (1602–1652), the Great Pyramid, and Modern Metrology." *Journal of the History of Ideas* 63, no. 4 (2002): 555–75.

Shapin, Steven, and Simon Schaffer, *Leviathan and the Air Pump: Hobbes, Boyle, and the Experimental Life.* Princeton, NJ: Princeton University Press, 1985.

Shapiro, Barbara. *John Wilkins 1614–1672: An Intellectual Biography*. Berkeley and Los Angeles: University of California Press, 1969.

Shulvass, M. A. "Sfarim vesifriot bekerv yehude Italia barenesans." *Talpiot* 4 (1910): 591–605.

Sinisgalli, Rocco, and Salvatore Vastola. *La Rappresentasione degli orologi solari di Federico Commandino.* Florence: Cadmo, 1994.

Sivin, Nathan. "Copernicus in China." *Studia Copernicana* 6 (1973): 63–122.

Sonne, I. "Book Lists through Three Centuries: A First Half of the Fifteenth Century, Italy." *Studies in Bibliography and Booklore* 2 (1955): 3–19.

Soykut, Muṣṭafā. "The Development of the Image 'Turk' in Italy through Della Letteratura dè Turchi of Giambattista Donà." *Journal of Mediterranean Studies* 9, no. 2 (2000): 175–203.

———. *Image of the "Turk" in Italy: A History of the "Other" in Early Modern Europe: 1453–1683.* Berlin: Klaus Schwarz Verlag, 2001.

Swerdlow, Noel. "The Derivation and First Draft of Copernicus's Planetary Theory: A Translation of the *Commentariolus* with Commentary." *Proceedings of the American Philosophical Society* 117 (1973): 423–512.

Tana, Li. *Nguyen Cochinchina: Southern Vietnam in the Seventeenth and Eighteenth Centuries.* Cornell University, Southeast Asia Program Publications, Studies on Southeast Asia, no. 23. Ithaca, NY: Cornell University, 1998.

Tekeli, Sevim. "Nasiruddin Takiyüddin ve Tyche Brahenin rasat aletlerinin mukayesesi (Taqī al-Dīn and Tycho Brahe's Observational Instruments)." Ph.D. diss., Istanbul University, 1958.

———. *16'ıncı asırda Osmanlılarda saat ve Takiyüddin'in Mekanik saat konstrüksüyonuna dair en parlak yıldızlar adlı eseri: The Clocks in Ottoman Empire in 16th Century and Taqī al-Dīn's "The Brightest Stars for the Construction of the Mechanical Clock."* Ankara: Ankara Universitesi Basimevi, 1966.

Terzioğlu, Derin. "Sufi and Dissident in the Ottoman Empire: Niyāzī-i miṣrī." Ph.D. diss., Harvard University, 1999.

Thoren, Victor E. *The Lord of Uraniborg: A Biography of Tycho Brahe.* Cambridge: Cambridge University Press, 1990.

Tietze, Andreas. *Muṣṭafā 'Ālī's Counsel for the Sultans of 1581.* 2 vols. Vienna: Verlag der Österreichischen Akademie der Weissenschaften, 1979.

———. *Muṣṭafā 'Ālī's Description of Cairo of 1599.* Vienna: Verlag der Österreichischen Akademie der Weissenschaften, 1975.

Tongas, Gèrard. *Les relations de la France avec l'empire ottoman durant la première moitiè du xviie siècle et l'ambassade à Constantinople Philippe de Harlay, comte de Cèsy, 1619–1640.* Toulouse: Impr. F. Boisseau, 1942.

Toomer, G. J. *Eastern Wisdome and Learning: The Study of Arabic in Seventeenth Century England.* Oxford: Clarendon Press, 1996.

Vela, C. "Luigi Cassola e il madrigale cinquecentesco." *Bollettino Storico Piacentino* 79 (1984): 184–88.

Vitestam, G. "As-Sidra(-t) al-muntahā." In *Actes du Premier Congrès International de Linguistique Sémitique et Chamito-Sémitique: Paris, 16–19 juillet 1969 réunis par André Caquot et David Cohen.* The Hague: Mouton, 1974.

Walbridge, John. *The Science of Mystic Lights: Qutb al-Din Shirazi and the Illumina-*

tionist Tradition in Islamic Philosophy. Cambridge, MA: Harvard University Press, 1992.

Westman, Robert. "The Astronomer's Role in the Sixteenth Century: A Preliminary Study." *History of Science* 18 (1980): 105–47.

———. "The Copernicans and the Churches." In *God and Nature: Historical Essays on the Encounter between Christianity and Science*, edited by David Lindberg and Ronald Numbers. Berkeley and Los Angeles: University of California Press, 1984.

———. "The Melanchthon Circle, Rheticus, and the Wittenberg Interpretation of the Copernican Theory." *Isis* 66 (1975): 164–83.

———. "Two Cultures or One? A Second Look at Kuhn's *The Copernican Revolution*." *Isis* 85, no. 1 (1994): 79–115.

Wilson, Arnold. "History of the Mission of the Fathers of the Society of Jesus, Established in Persia by the Rev. Father Alexander of Rhodes." *Bulletin of the School of Oriental Studies* 3, no. 4 (1925): 675–706.

Wilson, Curtis. *Astronomy from Kepler to Newton*. London: Variorum, 1989.

———. "Kepler's Derivation and Elliptical Path." *Isis* 59, no. 1 (1968): 4–25.

Y'ari, Avraham. *Hadfus ha'Ivri beQushta*. Jerusalem: Hebrew University Press, 1967.

Yates, Frances. *Giordano Bruno and the Hermetic Tradition*. Chicago: University of Chicago Press, 1964.

———. *Rosicrucian Enlightenment*. London: Routledge, 1975.

INDEX

209n15; of clocks, 15–16; and Galileo, 70; and print *vs.* scribal culture, 166; and Richelieu, 143, 144, 210n20; and Taqī al-Dīn, 46
Gilbert, William, 131, 204n78
Gingerich, Owen, 110, 140, 167
Giustiniani, Agostino, 62
Goa, 6, 47, 53, 55, 65, 74; Inquisition in, 60–61
Godinho, Manuel, 185n25
Greaves, John: and ancient manuscripts, 104, 117, 135, 138; associates of, 117–18, 128–35, 137; and astronomical nomenclature, 106–10; and astronomy, 104–6, 117, 128, 138; on Copernican system, 110–13, 127, 200n15; and English Civil War, 132–33; and Laud, 117, 120–22, 128, 137; on measurement, 106, 113–16, 137, 199n2; and Newton, 135–37; and print culture, 166; and Tycho, 105, 109, 126, 127; and universal language, 133, 135; and Ussher, 118, 122–28, 137; and Wilkins, 131–36, 138
Greece, 2, 166, 172n4, 207n9
Greek language: and astronomical nomenclature, 107; Bible in, 62–63, 125; and Greaves, 116, 117, 127; and James I, 118, 119; for "mirror," 146; translations from and to, 2, 23, 26, 27; and water clocks, 19
Greek Orthodox Church, 119–20, 121, 164

Habsburgs, 119, 143, 209n14; and Ottoman Empire, 28–31, 32, 33, 145, 181n91
Ḥādīth, 19
Hagecius, 30, 180n88, 181n94
HaLevi, Jacob, 22, 23, 97, 101–2, 177n56
Halley, Edmund, 1, 160
Harmonices mundi (Kepler), 120
Haskala (Jewish enlightenment), 77
al-Haytham, Ibn (Alhazen), 149
Hebraism, 61–67
Hebrew language: and ancient knowledge, 66, 73; ancient manuscripts in, 78, 164, 166; and apocalypse, 19; Bible in, 62–63, 74, 118, 121, 125; and Book of Job, 61, 63–65; and Greaves, 114, 116; and Hebraist studies, 62; and print culture, 62, 88, 103, 164; and Pythagoreanism, 67; and *Sefer Elim*, 76, 90; students of, 56, 74, 94, 130; translations from and to, 2, 5–6, 22, 23, 88; and universal language, 135
Heidegger, Martin, 3
Hermes, 89, 114, 147, 150
Hill, Donald, 19
Hinduism, 60, 66
Hipparchus, 85, 199n2
Historia dei riti hebraici (History of the Rites of Present Jews; da Modena), 98
Hobbes, Thomas, 105, 133
Holy Land, 6, 41; ancient texts in, 70, 73; della Valle in, 54, 55, 70, 74; and Karaites, 77, 86, 94, 96. *See also* Near East
The Hope of Israel (Ben-Israel), 96
Hulagu Khan, 27
Hveen Observatory, 28, 34, 120

Ibn Al-ʿArabī, Abū ʿAbdullāh, 150
Ibn-ʿEzra, Avraham (Avenezra), 86, 102, 125, 142, 187nn59, 60, 194nn40, 41, 207n10; on Book of Job, 64; and Copernican system, 88–89; on Elim, 90
Ibn-Maʿārūf, Taqī al-Dīn Muḥammad, 8–10, 101; and comet of 1577, 42; and David Ben-Shushan, 21–24; foreign sources of, 18–19, 20–21, 43; and *idrāk*, 150; intellectual sources of, 10–27, 31, 46; and mechanical clocks, 15–21, 38, 45, 46, 150; and millennialism, 45–46; and Ottoman Observatory, 34, 38, 40, 41, 43, 45; and Tycho, 8–9, 46
Ibn-Shāṭir, ʿAla al-Dīn, 20, 150, 156
Ibn Tufayl, Abū Bakr (Abubachar), 177n54
idrāk. *See* illuminist apperception
İhsanoğlu, Ekmeleddin, 155, 206n2
illuminist apperception (*idrāk*), 142, 147–53, 161, 213n49; and al-Zigetvari, 154, 159, 160; and Islamic astronomy, 162, 165, 213n48
The Incoherence of Philosophers (*Tahāfut al-falāsifah*; al-Ghazzālī), 161
India, 57, 60, 65, 67. *See also* Goa
In Job commentaria (de Zuñiga), 61, 65
Inquisition: and ancient versions of Bible, 65; and Campanella, 68, 72; and Copernican system, 59–61, 65, 67–68; in Italy, 21, 23; and print culture, 216n2
Iobus: Carmen Heroicum (Vavasseur), 63
Iskandrani, Jacob, 81–82, 84, 100, 192n13
Islam: apocalypse in, 19–20, 165; and astronomy, 8, 9, 35, 70, 82–84, 97, 106, 108, 135, 147, 154–55, 159, 160–62, 165, 213n48; converts from, 53, 54; converts to, 144; and Copernican system, 21, 172n4, 206n2; culture of, 1–7, 9, 30, 41–42, 45–46, 163–66; European criticism of, 30, 114; medical